ERRATUM

The contents of Table I on p. 206 should be replaced
by the contents of Table II on p. 222 and vice versa.

MODERN OXIDE MATERIALS

Preparation, Properties and Device Applications

MODERN OXIDE MATERIALS

Preparation, Properties and Device Applications

Edited by

B. COCKAYNE

Royal Radar Establishment, Malvern, England

and

D. W. JONES

Centre for Materials Science,
University of Birmingham, England

1972

ACADEMIC PRESS · London and New York

ACADEMIC PRESS INC. (LONDON) LTD
24/28 Oval Road,
London NW1

U.S. Edition published by
ACADEMIC PRESS INC.
111 Fifth Avenue,
New York, New York 10003

Library of Congress Catalog Card Number: 73–187923
ISBN: 0 12 177750 2

Printed in Great Britain
by Unwin Brothers Limited, The Gresham Press, Old Woking, Surrey, England
A member of the Staples Printing Group

PREFACE

Traditionally, oxides have been used as refractories and glasses but in recent years their basic properties have been further exploited in other fields such as optics and electronics.   This volume contains edited lectures presented to a Spring School (held at the University of Birmingham, U.K., April 1971) during which these new applications, and recent developments in the traditional use of oxides were extensively studied.   Major emphasis is directed towards the materials science and physical aspects, but for completion some comments on the chemical production of oxide materials are also included.

B. Cockayne

D.W. Jones.

Acknowledgement

The Editors wish to record their appreciation to Mrs. J. Heathcote for assistance in the organisation of the School, but particularly for carrying out the arduous task of typing the manuscript.

CONTENTS

OXIDES AS CRYSTALLINE LASERS

B. COCKAYNE

Royal Radar Establishment,
Malvern, Worcestershire

I.   INTRODUCTION

Crystalline oxide materials have played an important role in laser development. For instance, pulsed laser action at optical frequencies was first demonstrated in ruby (Al_2O_3/Cr^{3+}) by Maiman (1960) just over a decade ago, and

I

room temperature operation of a continuously emitting solid
state laser was first reported in the mixed oxide, calcium
tungstate ($CaWO_4/Nd^{3+}$), by Johnson et al. (1962).   Since
then, laser action has been demonstrated in a wide variety of
crystalline materials including many other oxides and some
semiconductors.   Parallel development has also taken place
using liquid and gaseous media.   Table I lists the more
common types of laser commercially available, amongst which
oxide materials figure prominently.

TABLE I

LASER MEDIUM	CLASSIFICATION	OPERATING WAVELENGTH (μm)
Al_2O_3/Cr^{3+}	Crystalline Oxide	0.694
$CaWO_4/Nd^{3+}$	Crystalline Oxide	1.06
$Y_3Al_5O_{12}/Nd^{3+}$	Crystalline Oxide	1.06
$Glass/Nd^{3+}$	Amorphous Oxide	1.06
GaAs	Semiconductor	0.84
He-Ne	Gas	0.633
A	Gas	0.488
CO_2-N_2-He	Gas	10.6

In the present article the principles and methods of
laser operation are outlined and the properties which make
oxides useful in the laser field are examined.   An account
of the preparation of laser crystals is included and applications for oxide lasers are briefly reviewed.

II.   LASER OPERATION

A.   STIMULATED EMISSION
A laser is an amplifier of light which produces gain in
the visible or near infrared region of the electromagnetic

spectrum.   It acts as a highly intense and coherent source
of radient energy.   The term 'laser' is the light wave
analogue of 'maser', an acronym for the phrase "microwave
amplification by the stimulated emission of radiation".
The extension of the maser principle to optical frequencies
was first proposed by Schawlow and Townes (1958).

Materials which are capable of stimulated emission
absorb energy in such a way that a population inversion is
produced between two available energy states, and then emit
the energy from the inverted state in the form of radiation
within a narrow wavelength range.   A three- or four-level
energy system is normally used to derive these states in the
visible and near infrared regions of the spectrum in order
to enhance inversion.   Energy systems of this type are
found in atomic, molecular and electronic states and these
form the respective bases of crystalline, gas and semiconductor lasers.

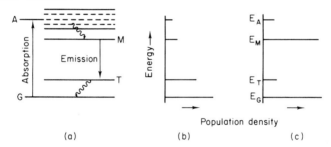

Fig.1.   A 4-level energy system
a) Transitions.   b) Boltzmann distribution.
c) Inverted distribution.

Solid-state crystalline lasers based on oxides use
atomic states, and a typical four-level energy system is
illustrated in Fig.1.   In the absence of excitation, the

population of the levels follows the normal Boltzmann distribution with the population of the lower levels exceeding
that of the higher levels.   If excitation is applied
selectively so that a quantum of energy, E_A-E_G, is absorbed,
the population levels are modified and the system is raised
into an unstable higher energy state, A, from which it decays
into the original energy state, G, the ground state, by
releasing energy in a series of steps from A to M, M to T
and T to G.   The possibility of obtaining stimulated
emission depends upon the rate at which energy is released
in these steps.   Under conditions where decay from the
absorption level, A, to the metastable level, M, is fast
compared with the following step to the terminal level, T,
the population in M builds up.   If at the same time decay
from T to G is also rapid, level M becomes more densely
populated than level T and an inverted energy state is
created between M and T compared to the normal Boltzmann
distribution.

     In the absence of any external influence, particles in
level M decay in a random manner to T, spontaneously emitting
radiation, and giving rise to the process known as fluorescence.   The frequency of the radiation emitted, ν_{MT}, is
defined by the expression

$$E_M - E_T = h\nu_{MT}$$

where h is Plank's constant and E_M and E_T are the respective
energies of the metastable and terminal states.   The transitions from A to M and T to G also emit energy but this occurs
by thermal vibration to the lattice.

     An alternative process to spontaneous fluorescence can
occur if radiation of frequency ν_{MT} is introduced into the

energy system while it is in an inverted state.   The
spontaneous decay from M to T is replaced by a stimulated
decay in which the stimulating radiation, ν_{MT}, interacts
with the excited energy system by forcing it to drop to T
and suddenly release the energy equal to $E_M - E_T$.   By definition the forced emission has the frequency ν_{MT}, thus
augmenting the stimulating radiation, and producing amplification.   In common with other forced oscillatory systems,
the stimulated emission is precisely in phase with the
stimulating radiation at the resonant condition.

Stimulated emission is, of course, the reverse process
to absorption.   If the population in level T exceeds that in
level M, so that the inverted condition is not realised, the
introduction of radiation with a frequency of ν_{MT} will
merely excite atoms from T to M giving an entirely absorptive
transition.

If the levels T and G coincide or if the temperature
of operation is such that level T is heavily populated by
thermal excitation, three-level operation results.   The
basic absorption and emission processes remain unchanged in
a three-level system but more than half of the ions in the
terminal level must be excited to the metastable level in
order to achieve the inversion necessary for stimulated
emission.

B.   RADIATORS AND RESONATORS

Light sources are not generally coherent since
individual atoms, molecules or electrons radiate spontaneously
in the visible region of the spectrum.   Stimulated emission
is a way of making the individual radiating particles work
together in a co-operative manner to establish coherency.

Three basic conditions are necessary for this state to be achieved. Firstly, a suitable system of energy levels must be found, secondly, an emissive condition must be established within the system and thirdly, the system must be contained within an optical arrangement which makes a large number of individual radiators emit in phase.

In solid state crystalline lasers, energy level systems capable of producing stimulated emission within the wavelength range 0.5 - 3.0 μm have been found amongst rare-earth ions (Nd^{3+}, Pr^{3+}, Ho^{3+}, Er^{3+}, Tm^{3+}, Yb^{3+}, Yb^{2+}, Sm^{2+}, Dy^{2+}), the transition metal ion (Cr^{3+}) and the actinide ion (U^{3+}). Cr^{3+} and Nd^{3+} are respective examples of three- and four-level systems. The active ion is contained in dilute solid solution within a crystalline host material, the preparation of which is described in a later section.

The extensive use of rare-earth energy states is not an accidental choice. The states are derived from the unpaired electrons of the 4f shell which are well shielded by two 5s and six 5p electrons from perturbations due to the crystal field of the host material. Hence, for any given ion, the basic positions of the energy levels are similar to the free ion and are approximately independent of the host material. Despite this shielding, some splitting of the energy levels does occur in the crystal field of the host and the nature of this splitting produces variations in laser performance from one material to another. The actinide ion is similar to the rare-earths, being derived from unpaired 5f electrons which are partially shielded by 6s and 6p electrons. The shielding is less effective than in the 4f series so the nature of the host assumes greater importance. The effect of the host material is most apparent for

transition metal ions where the usable energy states are
derived from the almost unshielded 3d electrons. The nature
and position of these levels are markedly different from the
free ion case and are so dependent upon the crystalline field
that the chance of ion and host combining to give a suitable
energy system is very small, hence the restricted number of
transition metal ions useful for stimulated emission. The
usefulness of the actinide series is restricted by the lack
of stable elements.

When the ions are contained in a suitable host, they
are excited into the emissive condition by the absorption of
energy from a lamp, frequently referred to as an optical pump,
which emits radiation over a broad range of wavelengths within
the visible and near infrared. The ions being excited must
have absorption bands within this wavelength range for
successful pumping to be achieved. The trivalent rare-earth
and actinide ions have narrow absorption bands whilst the
divalent rare-earth and transition metal ions absorb strongly
in this region of the spectrum.

Once the ions have been pumped into an inverted state,
the emissive condition is fulfilled and light amplification
can be produced by passing a beam of light of the correct
stimulating frequency into the excited crystal. The beam
of light stimulates the emission of more light of the same
frequency, thus producing optical gain.

It is more common practice to use the crystal as an
active medium which is made to oscillate inside a resonator
system. One such resonator is made by polishing plane
parallel surfaces on opposite ends of a crystal and applying
mirror coatings to both ends. An emissive condition can be
established by optically pumping the crystal through its

side.   Initially, pumping generates fluorescent rays
travelling in all directions.   However, rays which start to
travel along the crystal axis perpendicular to the mirrors
are continually reflected and amplified as they stimulate
emission from excited atoms in their path, whilst rays
travelling at an angle to the axis are lost through the sides
of the crystal, as illustrated in Fig.2.   Thus, optical gain
occurs by oscillation.   If one of the mirrors has a slightly
lower reflectivity than the other, coherent laser radiation
can emerge.

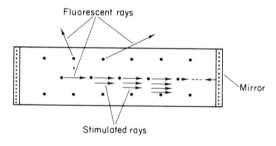

Fig.2.   Stimulation by oscillation.

A steady oscillation of this type can only be built up
provided that the gain due to stimulation is greater than
the losses due to factors such as absorption or transmission
by the mirrors, and scattering within the crystal.   If the
resonator has a length (L), the intensity of the radiation
(I) after a single complete pass of 2L along the resonator
can be defined in terms of the original intensity (I_0) by the
equation

$$I = I_0 R_1 R_2 \exp (\alpha .2L) . \exp (-\delta .2L)$$

where α and δ are the respective gain and loss coefficients
and R_1 and R_2 are the reflectivities of the two mirrors.   At
the steady state condition, which is referred to as the laser

threshold, $I = I_0$ and

$$R_1 R_2 \exp (\alpha - \delta) \cdot 2L = 1.$$

Above this threshold condition, the stimulated emission builds up to a level which is controlled by the intensity of the pumping. The factors which control α and δ are discussed more fully in later sections.

C.   LASER CAVITIES

A solid state crystalline laser medium is usually pre-pared as a circular sectioned cylindrical rod with polished sides and plane parallel ends. Typical sizes vary from 0.3cm diameter by 5cm long to 1.5cm diameter by 15cm long. The laser medium and pump lamp are normally placed in some form of focussing cavity in order to couple the output as efficiently as possible into the laser material.

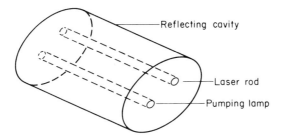

Reflecting cavity

Laser rod

Pumping lamp

Fig.3. An elliptical cavity.

Early cavities consisted simply of a helical flash tube with the laser rod positioned along the axis of the helix. Since then, a number of cavities have evolved. The most commonly used is an elliptical cylinder of the type illustrated in Fig.3 with the lamp and rod positioned along the focal axes of the cylinder. Spherical and ellipsoidal cavities have also been employed but these are more costly

and inconvenient to use than the elliptical system.   A
double elliptical cylinder allows the use of two pump lamps
which aid higher power applications.   A detailed description
of cavity geometries is provided by Evtuhov and Neeland
(1966).

Several different lamps have been developed for optical
pumping but the principal types are the tungsten-iodine filament lamp and rare-gas arc lamps.   Only a small fraction of
the lamp output is usefully absorbed by the laser medium.
A large proportion is dissipated as heat which produces
optical distortion.

D.   METHODS OF OPERATION

Solid state lasers can be operated in two basic ways.
These are pulsed and continuous wave (CW).   Pulsed operation is induced directly by single pulses from the flash
lamp.   CW operation is possible when the energy required to
induce laser action is small.   Under these circumstances a
continuously operating lamp can be employed and a continuous
beam of coherent radiation generated.

In either CW or pulsed operation, a laser system can
be Q-switched to produce large energy pulses.   This type of
operation occurs when the reflectivity of the output mirror
is reduced to zero for short periods of time whilst energy
is still pumped into the laser medium.   When the output
mirror is returned to normal reflectivity the stored energy
is suddenly released as a narrow high intensity pulse.   The
requisite change in reflectivity can be provided by rotating
the output mirror of the laser medium.   Mirrors external to
the laser medium must be used in this case instead of the
mirrored rod ends described earlier.

The normal laser rod can resonate in a number of ways both axially and transversely as the resonator dimensions are very much greater than the operative wavelength.   Only those wavelengths which are within the wavelength spread of the laser emission can resonate but normally there are enough of these to yield a multimoded output.   Some applications demand the generation of one particular resonance, that is single mode output.   Several methods of obtaining mode selection have been described by Geusic et al. (1970). Normally, the optical characteristics of the laser cavity are changed in some way so that all modes except the one required have a high loss factor.

III.   LASER MATERIALS

A.   THE PROPERTIES REQUIRED

The material ultimately used in a laser system depends to a certain extent upon the use of that system.   Some of the properties demanded of a material vary according to whether CW, pulsed, Q-switched, multimode or single mode operation is applicable, and according to the power output required.   For instance, a high gain material is favoured for CW operation in order to provide a low threshold but a lower gain is preferred for Q-switching in order to prevent the material lasing prior to the operation of the switch, and before the required amount of energy has been stored. However, the basic property requirements are common to most crystalline laser materials and these are summarised below -

Structural properties required
    1.   A substitutional site for the active ion
    2.   Single crystallinity

3. Mechanical hardness and stability

4. Chemical stability

Physical properties required

5. Low thermal expansion

6. High thermal conductivity

7. Low photoelastic constant

8. Low temperature coefficient of refractive index

Optical properties required

9. Optical transparency

10. Broad absorption bands

11. High quantum efficiency

12. Narrow linewidth

The prime requirement is that the host material will form a substitutional solid solution with the active ion at concentration levels which produce reasonable gain. Hence, materials with substitutional lattice sites of similar size and valency to the active ion are obvious choices. The fulfillment of this need must be coupled with the necessity to prepare the material, doped with the appropriate ion, in a state of good optical quality free from defects which cause the stimulated radiation either to be scattered or refracted away from the axis of the laser resonator. In most instances the requisite quality can only be obtained in single crystal material.

Certain properties are demanded by the processes used to fabricate a laser rod which involve cutting, grinding and polishing of the single crystal. The specification for the ends of a laser rod demand flatness to $\lambda/10$ using sodium light ($\lambda = 0.5898$ μm) and parallelism to $\pm$ 5 secs. of arc. This specification is only met readily by mechanically hard

materials.   Chemical stability is also important as the end
faces are usually subjected to a vacuum coating process in
which either metallic reflective layers (intra-cavity
mirrors) or dielectric anti-reflection layers (extra cavity
mirrors) are deposited.   Hence, crystals which either react
with the coating materials or decompose at low pressure are
excluded.

Chemical and mechanical stability have further relevance in relation to the environment of the laser rod.   In
the cavity, the rod has to be mounted firmly.   Consequently,
it is subject to differential thermal expansion during
optical pumping due to temperature gradients which develop
because a large amount of energy is dissipated as heat within
the rod.   The ensuing strains within the material can exceed
the yield strength resulting in failure by brittle fracture,
particularly in crystals having natural cleavage planes.
Materials with a low thermal expansion coefficient are preferred to minimise strain.   A high thermal conductivity is
also an advantage in assisting with heat removal.   This
latter process is enhanced by the use of liquid coolants.
These often contain a solute to absorb ultra violet radiation
emitted by the pumping source and thereby inhibit the common
problem of colour centre formation which adds to absorption
losses within the laser.   Hence, chemical compatibility
with the coolant is also an important criterion.

Most materials change refractive index with temperature
and with the stress induced by temperature changes.
Although these effects can be minimised by cavity design and
the use of coolants, it is obviously advantageous if the
temperature coefficient of refractive index and the photo-
elastic constant for the host are small.

Another fundamental requirement is that the host
crystals must be completely transparent at the absorbing and
fluorescing wavelengths of the active ion, which effectively
precludes compounds either exhibiting colour or possessing
absorption bands at infrared wavelengths less than 3 μm in
their undoped state.

For any given ion, the detailed lasing properties are
also determined by the host material.   The crystalline field
associated with the particular site where the active ion is
substituted splits the basic energy levels into a series of
sub-levels.   Properties such as absorption, quantum
efficiency and linewidth are modified by the nature of this
splitting and as the crystalline field varies from one host
and substitutional site to another, these properties vary
likewise.

Ideally, the host/ion combination should possess broad
absorption bands in order to inject as much as possible of
the available pumping radiation into the laser medium.
When the bands are very narrow, other species of ion are
sometimes added to the lattice to transfer optical energy
from their own absorption bands into those of the active ion.
This process must be non-radiative and was first applied to
laser materials by Kiss and Duncan (1964).   The quantum
efficiency of the host/ion compound should also be large,
which means that a high probability should exist for an
absorbed photon to result in the emission of a stimulated
photon.   In order to achieve a high gain, the output of the
lasing transition should have as small a spread in frequency
as possible, i.e. a narrow output linewidth.   Hence, gain
is greater with a single intense line than with a multi-
plicity of low intensity lines.   The energy level splitting

which controls these factors can also be changed by interactions between the active ions, particularly at high
concentrations.   In general, concentration levels are kept
below the interaction level.

B.   MATERIALS UTILISED

Inevitably, no material has all the requisite properties and the quest for the optimum compromise has led to the
evolution of a large number of host lattice/active ion
permutations.

TABLE II

CLASSIFICATION	COMPOUNDS
Oxide	Al_2O_3, Y_2O_3, Gd_2O_3, La_2O_3,
Oxide-Oxide	$Y_3Al_5O_{12}$, $Lu_3Al_5O_{12}$, $Y_3Ga_5O_{12}$, $Gd_3Ga_5O_{12}$, $YAlO_3$, YVO_4, $CaWO_4$, $CaMoO_4$, $SrMoO_4$, $PbMoO_4$, $NaGd(WO_4)_2$, $NaLa(MoO_4)_2$, $CaNb_2O_6$, $LiNbO_3$,
Fluoride	CaF_2, SrF_2, BaF_2, LaF_3, CeF_3, HoF_3,
Fluoride-Fluoride	$YLiF_4$, $NaCaYF_6$,
Oxide-Fluoride	$Ca_5(PO_4)_3F$,
Oxide-Sulphide	La_2O_2S

The restrictions on transmission and hardness limit
the useful materials mainly to oxide and fluoride based
compounds and their derivatives.   The limitations on
absorption restrict the cations of the host to those which
do not form coloured salts.   These conditions still leave a

formidable list of potential host cations to be used in
compound form with either oxygen or fluorine.   However, the
list is further restricted by the solid solubility condition.
For trivalent dopants, this criterion confines the compounds
to those containing either divalent cation sites with a size
similar to the active ion or trivalent cationic sites.
Similar criteria apply to divalent dopants.   These are
general conditions and exceptions can be found but they
illustrate the criteria which limit the presently known
laser host lattices to the type of compound categorised in
Table II.

In order to effect a comparison between the various
materials, the basic properties of the more important
compounds are given in Table III.   With the exception of
Cr^{3+} in Al_2O_3 (ruby), the laser properties are compared with
reference to the Nd^{3+} ion.   This choice is made because the
Nd^{3+} ion has dominated the crystalline laser field due to
its unique capability of maintaining high power room temperature operation.   Such operation is possible because the
terminal and ground energy levels of the ion are separated
sufficiently for the terminal level to be completely empty
at room temperature ($h\nu_{TG} > kT$ at 300 $^{\circ}K$).   The energy
required to produce inversion is therefore low and CW operation is readily achieved.   Ruby has attracted widespread
interest because it radiates coherently in the visible
spectrum.

The reasons for the importance of oxide materials as
solid state lasers are apparent from Table III.   One
distinctive feature is that the materials with high melting
points, which is indicative of strong bonding, have the
greatest mechanical stability, being extremely hard and free

TABLE III

MATERIAL PROPERTIES : HOST LATTICE	$Y_3Al_5O_{12}$	$YAlO_3$	Y_2O_3	$CaWO_4$	$Ca_5(PO_4)_3F$	CaF_2	La_2O_2S	Al_2O_3
Hardness (Mohs Scale)	8.5	8.5	6.8	4.5	5.0	4.0	5.5	9.0
Thermal Conductivity (mW/cm °C)	140	110	134	40	24	97	350	420
Expansion Coefficient (10^{-6}/°C)	6.9	2.2	7.0	11.2 (a) 18.7 (c)	8.5 (a) 9.1 (c)	19.5		6.8 (a) 5.7 (c)
Crystal Structure	Complex Cubic	Ortho-rhombic	Cubic	Tetrag.	Hexag.	Cubic	Hexag.	Hexag.
Site Symmetry	Tetrag.	1/m	2,$\bar{3}$	$\bar{4}$	m	Cubic* Rhombic Tetrag.		3mm
Cleavage Planes, etc.	None	Twins		{001}		{111}		None
Melting Point (°C)	1970	1875	2450	1566	1705	1360		2050
Crystal Production Technique	Czoch.	Czoch.	Verr.	Czoch.	Czoch.	Czoch. Brd/Stck.	Brd/Stck.	Czoch. Vern.
Distribution Coefficient (k,Nd^{3+})	0.2	0.62	0.2	0.2-0.8*	0.5	0.9		1.0 (Cr^{3+})
Lasing Wavelength (μm)	1.065	1.079	1.073	1.058	1.063	1.046	1.076	0.695
Linewidth at RT (cm^{-1})	6.5	8.0	10.0	35	6.0	120	8.0	
Relative quantum Efficiency (Emission/Absorption)	0.8			0.6				0.7
Fluorescent Lifetime (μsec.)	230	180	260	175	255	1100*	90	3000
Refractive Index Coefficient (10^{-6}/°C)	7.3							
	1,2,13	3	2,4	2,5,6	2,7	2	8,9	10,11,12

* Dependent upon the charge compensation mechanism for the 3+ ion.

Main References:-

Key: 1.Geusic et al.(1964); 2.Thornton et al.(1969); 3.Weber et al.(1969); 4.Hoskins and Soffer (1964); 5.Johnson et al.(1962); 6.Brandewie and Telk (1967); 7.Ohlmann et al.(1968); 8.Alves et al.(1970); 9.Wickersheim et al.(1970); 10.Maiman (1960),(1961); 11.Holland (1962); 12.Geller and Javorsky (1962); 13.Findlay and Goodwin (1970).

from fracture planes.   The outstanding materials are all
oxides or mixed oxides, the important examples being alumina
(Al_2O_3) and yttrium aluminium garnet ($Y_3Al_5O_{12}$).   Yttrium
orthoaluminate ($YAlO_3$) also has reasonable mechanical properties apart from the possible operation of twinning
mechanisms.   The strong bonding also confers a low thermal
expansion on these oxides.   The highest thermal conductivities are also displayed by this group of materials.   In
contrast, the fluorides and other compound groupings have
low thermal conductivities and poor mechanical properties,
exhibiting cleavage and being relatively soft.   Lanthanum
oxysulphide (La_2O_2S) is a possible exception.

Another important feature is that the materials which
provide a site of low symmetry for the substituting ion give
narrow linewidths corresponding to high gain.   Once more,
the oxides provide outstanding examples with $Y_3Al_5O_{12}$ and
$YAlO_3$.   The mixed oxide-fluoride known as fluorapatite
($Ca_5(PO_4)_3F$) and the mixed oxide-sulphide (La_2O_2S) are also
high gain materials. $Ca_5(PO_4)_3F$ and $YAlO_3$ are particularly
interesting in that the gain is crystal orientation
dependent and can be reduced from the values given by a
factor of 2-3 in some directions.   This allows some flexibility in the choice of gain for given applications.   For
example, high gain is required for CW operation and a
somewhat lower gain for Q-switching in order to increase the
amount of energy which can be stored.   The site symmetry
also influences the fluorescent lifetime which in turn also
controls the amount of energy which can be stored.   The
longer the lifetime, the greater the energy storage
potential.   Table III therefore clearly illustrates why
Cr^{3+}/Al_2O_3 is favoured for high powered pulsed systems whilst

Nd^{3+}/oxide systems are employed in either CW or low power
Q-switched operation.

All the types of material listed in Table III have the
requisite transmission properties.   Fluorides transmit
further into the infrared than the oxides and amongst the
oxides, materials such as calcium tungstate ($CaWO_4$) have
pronounced absorption bands in the 3-4 μm region.   However,
these differences in transmissivity have very little practical
significance as the absorption and emission spectra of the
presently known laser ions occur at shorter wavelengths than
3 μm.

Some of the basic properties such as quantum efficiency have only been measured on materials which have already
proved useful.   Hence, no direct comparison can be effected
between the various materials in these instances.   The
higher gain exhibited by oxide host lattices is reflected in
the lower theshold powers required to produce laser action.
However, properties such as threshold power depend on other
factors such as the active ion concentration, the state of
perfection of the laser rod and the method of testing, so
comparisons of this type are meaningful only in qualitative
rather than quantitative terms.   Hence, values for this type
of property are not quoted.

Some of the material properties also exert an influence
upon the ultimate state of perfection in which the crystalline material can be produced.   The hard isotropic lattices
can generally be grown as better quality single crystals
than either mechanically soft or anisotropic materials
because they are better able to withstand the thermal stresses
and resulting strains which develop to some extent during all
crystal growth processes.   This factor again favours the

high melting point oxide materials and particularly $Y_3Al_5O_{12}$.

C.   CRYSTAL GROWTH

Having established that oxide materials possess many
of the basic properties demanded of a solid state laser
material, there remains the problem of producing the material
as an optically acceptable single crystal. Several techniques have been employed in efforts to produce laser quality
material but only three have been developed for commercial
production. These are the Czochralski, Verneuil and
Bridgman-Stockbarger techniques and all involve the crystallisation of a crystal from its own melt as depicted in Fig.4.

In the Czochralski process, a single crystal seed is
dipped into a crucible containing the melt and then slowly
withdrawn. Melt solidifies on to the seed during the withdrawal process and the seed is grown into a crystal of the
required shape by adjusting the heat input into the melt in
order to control the crystal diameter. In the Verneuil
process, very finely divided powder is fed through an oxy-
hydrogen flame and is deposited on to the molten tip of a
single crystal seed which is gradually lowered away from the
hot zone as more material is deposited, thereby effecting
growth. The Bridgman-Stockbarger process employs a
crucible full of molten material which is steadily lowered
from a hot furnace into a cooler one. When the crucible
tip cools below the melting point, one or more crystals form
and continue to grow as more melt cools. The methods used
for particular materials are given in Table III.

The Czochralski technique is the most controllable and
thus far has been the most widely employed. It is, however,
limited by the availability of compatible crucible materials

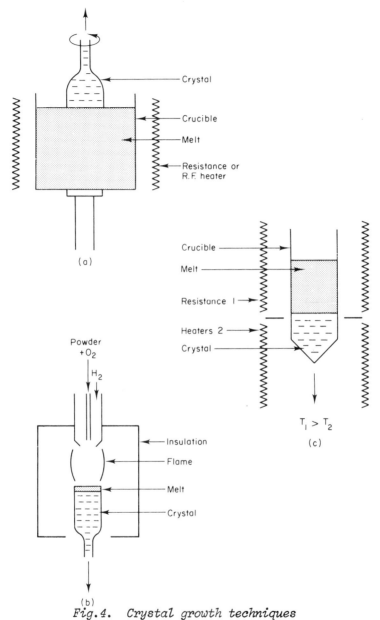

Fig.4. Crystal growth techniques
a) Czochralski.   b) Verneuil.   c) Bridgman-Stockbarger.

B. Cockayne

to compounds which melt below 2150°C. The Verneuil process
is characterised by steep temperature gradients which strain
the crystal, whilst the Bridgman-Stockbarger method is
limited by the difficulty of controlling crystal orientation
and by the lack of non-adherent/non-reactive container
materials at high temperatures.

Although oxides have the requisite basic properties
for laser hosts, they are generally difficult materials to
produce as perfect single crystals. The high melting point
leads to a number of technical difficulties in providing a
thermally, mechanically and chemically stable growth environment. Furthermore, as the only chemically compatible
crucible materials are the precious metals, platinum (used
up to 1500°C) and iridium (used up to 2150°C), the crystals
are expensive.

The methods of growth and associated problems have
been reviewed by Nassau (1964) and Cockayne (1968). One
problem of particular importance is that of substituting
sufficient active ions into the host lattice whilst maintaining optical quality. The substituting ion frequently
differs in size from the substituted site which produces an
uneven distribution of the active ion during crystal growth
because the ion dissolves to a different extent in the solid
than in the liquid. This is normally defined in terms of a
distribution coefficient, k (= concentration in solid/
concentration in liquid). Ions having a low k enhance
defect formation due to difficulties in rejecting the ions
in excess at the growing interface. This effect can only
be minimised by employing a low growth rate for the crystal
which increases the demands on environmental stability as
long time periods are then needed to produce a crystal.

The relevant values of k quoted in Table III show that the more favoured lattices from a laser viewpoint are the most difficult to dope with Nd^{3+}.

Despite these difficulties, in cases where the material has exhibited useful laser characteristics, technologies have been developed to solve the problems encountered.

D.   CRYSTAL DEFECTS

As a class of materials, oxides are little different from fluorides in the types of defect found in single crystals.   In laser materials, defects such as gas bubbles, voids and solid state precipitates scatter both incident and stimulated radiation thereby increasing the value of the loss coefficient, δ, and increasing the laser threshold. The formation of such defects is promoted by impurities and non-stoichiometry in the component materials, hence the controlled preparation of the components as described by Cooper et al. (1971) is an important factor in maintaining crystal quality.   Changes in refractive index within the material of only a few parts in 10^5 also exert a marked influence on laser performance by producing optical path differences between adjacent volumes of material which result in beam divergence, again increasing δ and threshold power. In anisotropic materials, the lattice tilts associated with dislocation low angle boundaries also cause changes in optical path.   Hence, cubic materials are preferred unless dislocation generation and movement is either difficult or can be controlled.   Refractive index changes are also induced by strain due to factors such as an inhomogeneous distribution of the active ion and thermal stressing during growth.

Lattice strains on an atomic scale can produce variations in the crystalline field from one ion to another and
thus cause a broadening of the fluorescent line which is
most readily observed at low temperatures. At room temperature, the dominating factor in line broadening is the
thermal vibration of the lattice. Vacant lattice sites and
trace quantities of impurities are also deleterious as they
often produce colour centres which act as an additional
absorption loss within the laser medium.

Most of these defects have to be controlled at the
growth stage as discussed by Cockayne (1968). Annealing
subsequent to growth does, however, assist in colour centre
removal and the relief of strain.

E.   CRYSTAL ASSESSMENT

The crystals produced by any of the methods available
are not optically perfect. Slight variations occur from
one crystal to another and sometimes within a given crystal.
Consequently, some assessment procedure is necessary in
order to select the best material. Particle scattering is
readily detected by a collimated light beam, whilst
refractive index changes can be observed by an interferometric examination. Both procedures are normally carried
out after plane parallel ends have been polished on the
crystal. In some instances, selection is made by examination of the laser emission pattern of the crystal at this
stage.

Although such tests act as a useful guide, stress
conditions within the material are altered by the cutting,
grinding and polishing procedures used to obtain the laser
rod, and by optical pumping. Hence the test which must

ultimately be applied is to make the finished laser rod emit within its working environment.

IV.   LASER APPLICATIONS

Oxide crystal lasers used for applications other than research have usually been based on Cr^{3+}/Al_2O_3 and $Nd^{3+}/Y_3Al_5O_{12}$. The predominant interest has been distance measurement because of the good transmission properties through clear air or space at the operating wavelengths. However, a variety of industrial applications have also evolved.   These have been fully described by Gagliano et al. (1969) and include welding, material removal and material shaping.   Welding has been confined to relatively thin materials such as the butt-jointing of 2mm thick steel plate and the joining of integrated circuits to thin film gold conductor paths on a ceramic substrate.   Material removal includes the drilling of holes in hard materials such as diamond dies and ruby bearings and the trimming of capacitors in thin film circuits.   Crystalline lasers are attractive propositions for work of this nature because the short pulse lengths available give a narrow heat affected zone and the small beam diameter permits a high degree of accuracy to be attained.   Furthermore, sufficient power is available to vaporise small volumes of material.   Thus far, material shaping has been restricted to brittle materials such as the scribing of silicon transistor wafers prior to isolation into individual units.

Ruby lasers have also found medical application in eye therapy whilst $Nd^{3+}/Y_3Al_5O_{12}$ has potential use in optical communications.   This latter application has been reviewed

fully by Geusic et al. (1970) and is feasible but requires
further development before it is economically competitive
with lower frequency systems currently in operation.

V.   SUMMARY

Oxide laser materials appear to be firmly established.
They exhibit a wide range of power capability. Nd^{3+}/
$Y_3Al_5O_{12}$ CW systems emitting 1100 W have been reported
whilst up to 50 kW is available in Q-switched operation.
Q-switched ruby has an output of up to 1000 MW.   Alternative
hosts for Cr^{3+} are unlikely because of the critical crystalline field conditions.   Possible alternatives for Nd^{3+}
again contain at least one oxide component but none have so
far matched the performance of $Y_3Al_5O_{12}$.   The poor thermal
conductivity of $Ca_5(PO_4)_3F$ limits its use in high power
applications and twinning in $YAlO_3$ is not yet fully controlled. La_2O_2S appears promising but has not so far been
fully evaluated. Nd^{3+}/glass is a possible alternative
since optical perfection is much easier to achieve.   However,
poor thermal conductivity will again preclude high power CW
use, although low power CW use may be possible.

A major shift in emphasis from oxide lasers is possible
if a change in active ion proves advantageous.   At the
moment this appears unlikely but the laser action reported in
HoF_3 by Devor et al. (1970) using Ho^{3+} transitions is
interesting because no problems of ion substitution arise.
However, this laser still has to be cooled to empty the
terminal state sufficiently for efficient laser operation.
The main competition in high power use is more likely to
arise from non-crystalline media such as the gas laser based

on CO_2-N_2-He where CW output powers of up to 8.8 kW have been reported.

REFERENCES

Alves, R.V., Wickersheim, K.A. and Buchanan, R.A. (1970) Bull. Amer. Phys. Soc. 15, 355.

Brandewie, R.A. and Telk, C.L. (1967) J. Opt. Soc. Amer. 57, 1221.

Cockayne, B. (1968) "Crystal Growth" (Frank et al., ed.) North Holland (Amsterdam) p.60.

Cooper, B.S., Bradley, N.J. and Hobbs, D.J. (1971).   This volume.

Devor, D.P., Soffer, B.H. and Robinson, M. (1970) Int. Conf. on Quant. Electronics (Kyoto).

Evtuhov, V. and Neeland, J.K. (1966) "Lasers" Vol.1 (E.K. Levine, ed.) Edward Arnold.

Findlay, D. and Goodwin, D.W. (1970) Adv. in Quant. Elec., 1, 70.

Gagliano, F.P., Lumley, R.M. and Watkins, L.S. (1969) Proc. I.E.E.E. 57, 114.

Geller, R.F. and Javorsky, P.J. (1945), J. Res. Nat. Bur. Standards (USA) 35, 395.

Geusic, J.E., Bridges, W.B. and Pankove, J.I. (1970) Proc. I.E.E.E. 58, 1419.

Geusic, J.E., Marcos, H.M. and van Uitert, L.G. (1964) Appl. Phys. Lett. 4, 182.

Holland, M.G. (1962), J. Appl. Phys. 33, 2910.

Hoskins, R.H. and Soffer, B.H. (1964) Appl. Phys. Lett. 4, 22.

Johnson, L.F., Boyd, G.D., Nassau, K. and Soden, R.R. (1962) Phys. Rev. 126, 1406.

Kiss, Z.J. and Duncan, R.C. (1964) Appl. Phys. Lett. 5, 200.

Maiman, T.H. (1960) Nature, 187, 493.

Maiman, T.H. (1961) Phys. Rev. 123, 1145.

Maiman, T.H. (1961) Phys. Rev. 123, 1151.

Nassau, K. (1964) Lapidary Journal 18, 42.

Nassau, K. (1964) Lapidary Journal 18, 386.

Nassau, K. (1964) Lapidary Journal 18, 474.

Ohlmann, R.C., Steinbruegge, K.B. and Mazelsky, R. (1968)
  Appl. Opt. 7, 905.

Schawlow, A.L. and Townes, C.H. (1958) Phys. Rev. 112, 1940.

Thornton, J.R., Fountain, W.D., Flint, G.W. and Crow, T.G.
  (1969) Appl. Opt. 8, 1087.

Weber, M.J., Bass, M., Adringa, K., Montchamp, R.R. and
  Comperchio, E. (1969) Appl. Phys. Lett. 15, 342.

Wickersheim, K.A., Buchanan, R.A., Alves, R.V. and Sobon, L.E.
  (1970) Bull. Amer. Phys. Soc. 15, 393.

OXIDES FOR ACOUSTO-OPTIC APPLICATIONS

M.V. HOBDEN

Royal Radar Establishment,
Malvern, Worcestershire

I.   INTRODUCTION

During the past decade there has been a rapid growth of interest in devices for controlling laser beams for a diversity of applications including optical stores, laser displays, communications, laser radars and laser Q-switching. This has stimulated research and development of better materials for these devices with properties tailored to the required performance.   In the last few years several new materials have been successfully produced for acousto-optic devices and some of these are oxides.

The objective of this article is to give the materials scientist an outline of recent developments and the devices that have evolved, and to show that there is still room for improvement, especially in optical quality and cost.

II.   ACOUSTO-OPTIC INTERACTIONS

The propagation of light in transparent materials is governed by the refractive index.   Internal strain within the material can modify the refractive index by the photo-elastic effect.   Periodic refractive index changes induced by externally generated strain waves can be used for modulation and deflection of light beams in acousto-optic devices by this acoustic-optic interaction.

The refractive index (n) of a material is related to the polarisability of the electrons in the lattice.   The more easily the electrons are shifted under the influence of the electromagnetic field of the light, the higher is the refractive index and the lower the wave velocity.   The most polarisable electrons are those least tightly bound by the electric fields within the lattice;   these are the valence electrons and those in orbitals of large radius.   It is the density and distribution of these weakly bound electrons that primarily determines the refractive index.

Strain alters the internal potentials of the lattice and this changes the shape and size of the orbitals of the weakly bound electrons causing changes in the polarisability and refractive index.   The effect of strain on the refractive indices of a crystal depends upon the direction of the strain axes and the direction of the optical polar-isation with respect to the crystal axes.   Consequently, the relationship between refractive index and strain is expressed by a photo-elastic tensor with several photo-elastic coefficients (p).   Crystals of all classes show photo-elastic effects and in the most general case there are thirty-six independent coefficients in the photo-elastic

tensor.   For crystals of higher symmetry there are fewer
coefficients, and in isotropic media such as glasses there
are two (designated p_{11} and p_{12}).   Photo-elastic
coefficients are dimensionless, and in practice are usually
in the range - 0.2 to + 0.5.

The magnitude of the change in refractive index can be
understood with the help of a simple example.   Consider a
longitudinal strain ε_{11} on an isotropic material.   For light
propagating perpendicular to the strain axis, the changes in
refractive index for polarisations parallel and perpendicular
to the strain axis are given by

$$\Delta n_{||} = -\frac{n^3}{2} p_{11}\, \varepsilon_{11} \quad \text{and} \quad \Delta n_{\perp} = -\frac{n^3}{2} p_{12}\, \varepsilon_{11}$$

The change of refractive index for unit strain is seen to be
proportional to $n^3 p$.   Refractive indices are typically ~ 2
and the maximum strain that can be applied before irreversible
damage occurs is $\sim 10^{-4}$, so the maximum changes in refractive
index that can be induced by strain are $\sim 10^{-4}$.   In most
practical applications the changes of refractive index are
typically $\sim 10^{-5}$.

To use these small changes to advantage it is necessary
to devise techniques where the optical effects depend upon
the differences rather than the absolute magnitude of the
refractive indices.   One method is to set up the strain as
a periodic wave in the material by means of an electrically
driven piezoelectric transducer bonded to a surface.   If
the wavelength (Λ) of this strain wave is comparable to the
wavelength of the light (λ) then the periodic refractive
indices behave as a three dimensional optical diffraction
grating and can be used to diffract light.

In Fig.1 the alternating layers of high and low refractive index retard and advance the wave to give a corrugated wavefront at the exit surface.   This wavefront resolves into plane waves, $PW_{\pm m}$, diffracted through angles $\theta_m \simeq m\lambda/\Lambda$ where m is an integer.   As the acoustic wavelength $\Lambda = v/f$, (where v is the velocity of sound in the material, and f is the frequency), it can be seen that θ_m is a function of frequency.   Changing the frequency of the acoustic wave is the basis of the acousto-optic beam deflector.

The phase change introduced by the strain wave has a maximum value $2\pi L.\Delta n/\lambda$.   The relative intensity of the diffracted waves is proportional to the square of this phase change and so, for a given strain, is proportional to $n^6 p^2 L^2/\lambda^2$.   For the purposes of comparing the diffracting efficiency of one material with another it is more realistic to compare them in terms of equal acoustic power per unit area (P*) than equal strain.   When this is taken into account it can be shown that the diffracting efficiency (η) is given by

$$\eta = \frac{\pi^2}{2} \cdot \left(\frac{n^6 p^2}{\rho v^2} \right) \cdot \frac{L^2}{\lambda^2} \cdot P* \cdot$$

The parameter $M_2 \equiv (n^6 p^2/\rho v^3)$ depends upon the properties of the material and is referred to as the figure of merit. For efficient devices it is desirable to use materials with high figures of merit, provided of course that they satisfy the secondary conditions of low optical and acoustic losses, and are available with sufficiently good optical homogeneity.

The figure of merit (M_2) is a measure of the efficiency of diffraction for a material in a device of fixed length and optical wavelength. When the design of acousto-optical devices is considered in more detail other figures of merit can be defined which are more appropriate. In the next section it will be shown that for Bragg diffraction modulators a more appropriate figure of merit is $M_3 \equiv (n^7 p^2/\rho\ v^2)$. It can also be shown that at very high acoustic frequencies the limitations of acoustic power density call for the use of $M_4 \equiv (n^8 p^2 v/\rho)$, and in certain optical deflection devices the figure of merit $M_1 \equiv (n^7 p^2/\rho v)$ is more suitable.

III. ACOUSTO-OPTIC MATERIALS

There are four important factors governing the choice of material for acousto-optic applications. Firstly, the appropriate figure of merit should be high enough to give an acceptable diffraction efficiency. Secondly, the acoustic and optic absorption must be sufficiently low. Thirdly, the optical homogeneity must be adequate. Fourthly, the material must be readily available at an acceptable cost.

It has been stated that there are four figures of merit for device materials M_1, M_2, M_3 and M_4 which apply to different device configurations and limitations. There are several values for a given figure of merit in one material according to the particular orientations of the light and the strain wave. In tabulated data it is usual to quote the highest figure of merit appropriate to the most effective arrangement for a given crystal and optical wavelength.

The search for new materials with good figures of merit is aided by published data on refractive indices and densities. The velocities of sound are known for some materials but the photo-elastic constants are quite rare.

In the absence of the necessary data there is a need for empirical rules to suggest which materials could be useful, followed by measurements on specimen crystals.

The various properties that appear in the figures of merit are related. For instance, refractive indices are usually greater in dense materials and the velocity of sound also depends on density. The situation is clarified by using the approximate relationship $v \simeq (c/\rho)^{\frac{1}{2}}$ where c is the relevant elastic modulus. This substitution gives the approximate expressions

$$M_1 = \frac{n^7 p^2}{\rho^{\frac{1}{2}} c^{\frac{1}{2}}} \quad : \quad M_2 = \frac{n^6 p^2 \rho^{\frac{1}{2}}}{c^{3/2}} \quad : \quad M_3 = \frac{n^7 p^2}{c} \quad : \quad M_4 = \frac{n^8 p^2 c^{\frac{1}{2}}}{\rho^{3/2}}$$

Other things being equal, a high refractive index is important. In practice, transparent materials can have a refractive index in the range 1.5 to 2.5 and this can intro-duce a factor of up to sixty in the figure of merit. For devices such as modulators that depend upon M_2 or M_3, a soft dense material has an advantage which does not conflict with a high refractive index. On the other hand, at the highest frequencies where M_4 can be applicable, hard materials of high refractive index but low density are more suitable. What sorts of materials fit into these categories?

The refractive index of a material is related to the polarisability of the electrons. For a high refractive index the material should have a high density of weakly bound electrons in large orbitals. When this is the case the material tends to have strong optical absorption in the near ultra-violet or even in the visible part of the spectrum due to transitions between electron states with small energy differences. The empirical expression

$$n^2 - 1 \simeq \frac{E_g}{15}$$

relates the energy gap (E_g, in eV), above which the material begins to absorb strongly, and the refractive index in the transparent part of the spectrum. It can be seen that for transparency in the visible $E_g \geq 3$ eV and $n \leq 2.5$ approximately. Ideally a material for use in the visible spectrum should have $n \simeq 2.5$. For use in the near infra-red it is possible to use low energy gap materials, for example the sulphides, with $n \simeq 3$.

Strongly ionic materials such as the alkali halides tend to have large energy gaps and fairly low refractive indices. LiF has a refractive index of about 1.4. When the anions are large and closely packed, there is a fairly high density of weakly bound electrons (those in the outer shell of the anion) and the refractive index is somewhat larger. LiI has a refractive index of 1.96.

On the other hand, strongly covalent materials such as Ge, Si or the III-V compounds tend to have too small an energy gap for use as transparent optical materials in the visible. They are suitable for some infra-red devices, and their very high refractive indices help to compensate for the decreasing diffraction efficiency with increase in optical wavelength. Some compromise has therefore to be made between strongly ionic and strongly covalent compounds. Oxide materials fall into such categories, the degree of ionicity or covalency depending upon the constituent atoms and type of structure.

If the refractive indices of the oxides of single elements are studied, it is found that there is a general trend from $n \simeq 1.5$ at low atomic number to ~ 2.4 at high

atomic number.   Superimposed upon this trend there are
large increases in refractive index in the regions of the
transition metals.   The oxides of the first transition
metal group tend to have n $\simeq$ 2.5 and be slightly coloured,
as would be expected because of the absorption energy gap.
For the second and third groups the data is somewhat scanty,
but in those cases where the oxides have been studied they
are often opaque.   The reason for this behaviour is the
existence of the d-shells of the transition metal cations
which tend to form a low lying conduction band or upper
energy level for the valence electrons, in the outer
orbitals of the oxygen.   These oxides are not generally
used in devices but there are very many mixed oxides having
a transition metal oxide as a constituent which have shown
promise.   In many cases they have E_g $\simeq$ 4 eV and n $\simeq$ 2.2.
Mixed oxides are the most promising oxide materials at the
present time for device work, and include $LiNbO_3$, $LiTaO_3$,
$PbMoO_4$ and the recently discovered Pb_2MoO_5 all with
refractive indices about 2.2 and transparent in the visible.

A disadvantage of materials containing transition
metal oxides is their tendency to lose oxygen during the
high temperature growth process because of the different
states of valency that they can take.   When the structure
is that of the oxide with the highest valence state, the
material is usually transparent because the oxygen valence
band is filled and the conduction band is empty.   When this
structure loses oxygen, excess cations tend to move to
interstitial sites and donate electrons to energy levels
from which they can be excited by visible light, so causing
optical absorption.   For this reason, many of these
materials need to be annealed at high temperatures in a

suitably pressurised oxygen atmosphere to render them transparent and free from defects.

The photo-elastic constants (p) cannot be accounted for by any simple model, as they depend upon the detailed electronic structure of the material. The strain in the material changes the potential within the lattice and the weakly bound electrons will respond by altering the size and shape of their orbitals. This causes the polarisability and refractive index to change. At the present time there is no reliable method of estimating photo-elastic constants. From the data that has been collected it seems that there is a tendency for the photo-elastic constants of the harder materials, such as the insoluble oxides and the alkali halides, to be smaller than those of the softer soluble oxides and molecular solids. There is considerable variation in these constants even in one material, and in practice they have to be determined experimentally.

The stiffness or elastic modulus (c) of a material can be related in a general way to the binding energy and melting point. Those with a high melting point have a high binding energy and are usually the stiffest. For modulators and beam deflectors a low stiffness is an advantage, and as this is usually connected with a low melting point it is very welcome. A low melting point usually means fewer growth or production problems, and a better chance of good optical quality. The softest oxides tend to be those with heavy metal cations especially lead, bismuth and thallium. The general picture of an ideal material emerges, somewhat indistinctly, as a dense, soft material of mixed bonding, probably an oxide and preferably a mixed oxide with a transition metal in its highest valence state as one

constituent, with an energy gap about 4 eV, and relatively
low melting point.   The exception is for the case where M_4
applies, i.e. when the frequency is very high and transducer
power density is limited; in this case the material should
have a high refractive index and be hard, yet low in density.
This could be for example an oxide with a light transition
metal atom such as TiO_2 or $MgTiO_3$.

The design of acousto-optical devices is also influenced by the acoustic absorption of the material.   In most
devices the drop in acoustic power across the aperture of
the device due to absorption has to be < 1 db for an
acceptable performance.   For large aperture beam deflectors
this can present a limiting factor in the overall performance.   The dominant absorption mechanism in solids is
attributed to anharmonic phonon effects in the lattice.
The strength of this absorption is not easily related to
other material properties though there is a tendency for low
losses to be associated with low density, strongly bonded
materials of high melting point.   For a given material the
acoustic absorption coefficient increases proportionally to
the square of the frequency; this tends to put a well
defined limit on the upper working frequency.   Very soft
infra-red transparent materials, such as As_2S_3, can be used
up to about 50 MHz, $PbMoO_4$ up to about 300 MHz, and $LiNbO_3$
up to 1 GHz.

The third factor influencing the choice of material is
the optical homogeneity, meaning the variation of the
refractive index from point to point due to inbuilt strains,
variable composition and other defects.   Variations of this
sort will affect the optical beams and degrade the resolution
of the device.   Most acousto-optical devices have an

optical path length in the region of a centimetre. This
implies that to keep aberrations of the wavefront to less
than $\lambda/10$, the fluctuations in refractive index must be less
than 5.10^{-6} over the whole aperture. Variations of refractive index as small as this over an aperture of several
centimetres (required, for instance, in high resolution beam
deflectors) are equivalent to the standards of homogeneity
of a good glass. It is no wonder that oxide single
crystals pulled from the melt rarely achieve this standard
of perfection. Single crystals of water soluble oxides are
less prone to optical inhomogeneity due to the slower growth
rate and the greater control of growth conditions.

The last and most important factor affecting the
development of acousto-optic devices is the material availability. No matter how good the published figures of merit,
if the material cannot be obtained as a single crystal of the
required size and quality it is of no practical use. It is
a matter of experience that work on most materials is eventually discontinued because of difficulties in preparation
and growth.

Table I summarises the essential properties of several
materials that might be considered for acousto-optic
applications. The figures of merit have been expressed in
terms of those of fused silica, which is taken as unity,
because it is a material commonly used in comparative
measurements. The data applies to longitudinal acoustic
waves and an optical wavelength $\lambda = 0.63$ μm.

Water has frequently been used for laboratory acousto-
optic work for many years. It has a large M_2 and M_3 in
spite of the low refractive index, because of the low
velocity of sound and high photo-elastic constant. It is

M. V. Hobden

an extremely convenient material in some respects due to the availability and lack of restriction on size or shape. Unfortunately, it has a very high acoustic loss which restricts the use to below about 30 MHz and also there is a difficulty in coupling the acoustic power into it which restricts the bandwidth to a small fraction of the working frequency.

TABLE I

	n	ρ_{g/cm^3}	$v_{10^5 cm/sec}$	M_1	M_2	M_3	M_4	Melting Point ^oC
LiTaO$_3$	2.18	7.4	6.2	1.4	0.9	1.4	2.3	1650
LiNbO$_3$	2.20	4.7	6.6	8.0	4.6	7.8	15	1253
Al$_2$O$_3$	1.76	4.0	11.1	0.9	0.2	0.5	3.9	2045
TiO$_2$	2.58	4.6	7.9	7.9	2.6	6.2	25	1750
YAG	1.83	4.2	8.6	0.1	0.1	0.1	0.3	1970
Pb$_2$MoO$_5$	2.18	7.1	2.3	31	84	63	11	950
α-HIO$_3$	2.20	4.6	2.4	14	55	32	3.6	d.110
PbMoO$_4$	2.38	6.9	3.7	15	24	25	10	1060
Fused Silica	1.46	2.2	5.9	1	1	1	1	1710
SF4 Glass	1.75	4.8	3.7	0.2	4.3	3.3	0.1	
H$_2$O	1.33	1.0	1.5	5.5	106	22	0.3	0

Fused silica has little to recommend it apart from the availability and very high quality which makes it suitable for devices where optical losses must be kept low, e.g. in intra-cavity laser modulators working up to about 300 MHz. Dense flint glasses such as SF4 fall into roughly the same category; they are somewhat more efficient but their acoustic attenuation limits them to below 100 MHz.

$PbMoO_4$ is typical of the oxide materials being developed for acousto-optic applications up to 500 MHz. It is dense and soft, has a high refractive index and relatively low velocity of sound. It is grown from the melt at about $1060^{o}C$ and has the scheelite type structure $(CaWO_4)$ with point group 4/m. It is unsoluble in water, chemically stable in air and fairly easy to fabricate into devices. It is not particularly easy to grow in very good optical quality. There is a tendency for the melt to lose PbO and oxygen, causing the stoichiometry to vary. The crystal is prone to dislocations and low angle boundaries and needs to be annealed in air or oxygen for a considerable period. Recently, $PbMoO_4$ has appeared on the market in considerably improved quality, but it is unlikely to achieve the optical quality of a reasonably good glass. It is also rather expensive for general use.

Another lead molybdate, Pb_2MoO_5 of point group 2/m, has been pulled from the melt at $950^{o}C$. While it is too early to predict the utility of this material it may well offer some improvements over the compound $PbMoO_4$ if it can be grown easily with good optical quality.

Alpha iodic acid $(\alpha-HIO_3)$ is an unusual example of an acousto-optic oxide material. This compound is essentially a molecular solid grown from aqueous solution, consisting of

HIO_3 molecules held together by Van der Waals forces and a small amount of hydrogen bonding. The high refractive index arises from the $I\bar{O}_3$ group which has some weakly bound electrons in large orbitals. The relatively large photo-elastic constants almost certainly arise from the ease with which the molecules can be forced together due to the weak binding. The principal disadvantages of this material are poor chemical and physical stability. The extremely high solubility in water makes it very susceptible to atmospheric humidity, and the relatively low strength of the material calls for special delicate methods for bonding the trans-ducers. While a very interesting and useful material for occasional laboratory devices, it is too expensive and difficult to handle for general use.

Yttrium aluminium garnet (YAG) has been included to show how totally unsuited to acousto-optic applications an oxide can be. By chance, the electron structure in this material is such that the photo-elastic effects are very small. This is one reason for the low strain birefringence of YAG grown for laser rods.

Rutile (TiO_2) and sapphire (Al_2O_3) are superficially similar in that they are both hard oxides with melting points in the region of $2000^{\circ}C$. Rutile has a much higher refractive index due to the fact that titanium is a transition metal and the energy gap is consequently much smaller. For acousto-optic device applications sapphire is virtually useless but rutile does have some advantages at the very highest frequencies where acoustic power density is important.

Lithium niobate and lithium tantalate have been developed primarily for non-linear optics and piezoelectric

applications.   The acousto-optic figures of merit are
fairly high, especially for the niobate.   Their principal
advantage lies in the low acoustic absorptions which allow
them to be used up to 1 GHz in certain applications.   The
large research effort put into these materials in the past
six years now enables crystals to be produced to a high
standard of optical perfection at reasonable cost.   There
are minor problems associated with these materials, such as
their susceptibility to optical refractive index changes due
to long term exposure to laser beams, and the sensitivity of
their refractive indices to temperature.

IV.   ACOUSTO-OPTIC DEVICES

The basic configuration for an acousto-optic device is
shown in Fig.1 where light is diffracted by the periodic
change in refractive index.   When $L \ll n\Lambda^2/2\pi\lambda$ and the
input acoustic power is low enough that $\Delta n < \lambda^2/n\Lambda^2$ the
device behaves as a two dimensional grating and light is
only diffracted into the first-order plane waves PW_{+1} leaving
the undiffracted light in PW_0.   This is a low powered,
inefficient device used sometimes in the research laboratory
because of the ease with which the performance can be
analysed.   As the power is significantly increased, higher
order beams PW_{+2}, etc. can be produced as well as $PW_{\pm1}$.
This type of diffraction is termed Raman-Nath diffraction.
Devices based on this configuration are inefficient because
of the restriction on the length of material over which the
light and sound can interact.   When $L \simeq n\Lambda^2/2\pi\lambda$ the
diffraction takes place into many of the waves $PW_{\pm m}$ but the
efficiency is correspondingly greater.   Devices using this

configuration are sometimes used to advantage in modulators
and laser Q-switches where the principal objective is to
reduce the intensity of PW_0 without the alignment problems
that arise in the Bragg angle modulator.

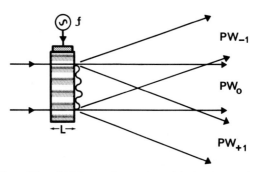

Fig.1. The diffraction of a light beam into
many plane waves by an acoustic strain wave.

When L >> $n\Lambda^2/2\pi\lambda$ diffraction occurs only when the
Bragg reflection condition is satisfied.   This is
illustrated in Fig.2 which shows diffraction taking place
into PW_1 when the input beam meets the acoustic wavefronts
at the grazing angle $\theta/2$, where sin $(\theta/2)$ = $\lambda/2\Lambda$.   Because
of the greater length over which diffraction takes place,
and the production of only one diffracted beam, this arrangement is used for high efficiency acousto-optic beam
deflectors.

The essentials of a simple modulator operating in the
Bragg configuration are shown in Fig.2.   Electrical drive
power at frequency (f) sets up a travelling acoustic wave of
wavelength (Λ), diffracting some of the incident beam into
PW_1 at the Bragg angle from the undiffracted beam PW_0.   The
relative intensity of PW_0 can be reduced by increasing the
acoustic power.   At very low powers the reduction in

intensity is directly proportional to the acoustic power.
At high drive powers the reduction of intensity is not a
linear function of the power but nevertheless it is possible
in principle to achieve 100% diffraction for a finite
acoustic power.

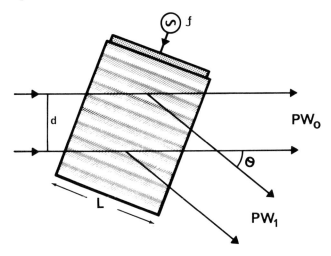

*Fig.2. Bragg diffraction of a light beam into
one plane wave by an acoustic strain wave.*

Any change in the acoustic wave takes a time $\tau = d/v$
to sweep across the optical beam.   This limits the switching
time to τ when it is used as an optical switch and limits the
bandwidth to $\Delta f = \frac{1}{\tau}$ when it is used as a modulator.   The
velocity of sound in most materials is about 5.10^5 cm/sec.
and so for a 10 MHz bandwidth, i.e. enough for one video
channel, d must be about 0.5mm.   It is relatively easy to
put one video channel onto a typical gas laser beam with
such a device.

For greater bandwidths the beam must be narrowed by
focussing.   However, as the beam becomes narrower it suffers

self-diffraction through a larger angle.   In order that the
Bragg condition can be met over the whole volume of the
device it is necessary that the diffraction of the acoustic
beam Λ/L should match the optical diffraction λ/nd.   That
is, the length should be chosen so that

$$L \; = \; \frac{nd\Lambda}{\lambda} \; = \; \frac{nv^2}{\lambda \, f \, \Delta f}.$$

It was shown earlier that the diffracting efficiency
of a device was proportional to $(n^6 p^2/\rho v^3) \, (L^2/\lambda^2)P^*$.
Substitution of the above value for L shows that for a given
frequency, bandwidth and power density, the diffraction is
proportional to $(n^8 p^2 v/\rho) \, (\lambda f \Delta f)^{-2} \, P^*$.   The basis of
comparison for materials in this situation is $M_4 \equiv (n^8 p^2 v/\rho)$.
It can likewise be shown that for optical beams of circular
cross section where total power is limited, the basis for
comparison is $M_3 = (n^7 p^2/\rho v^2)$.

Acousto-optic amplitude modulators can compare favourably with electro-optic modulators in cost of manufacture
and power requirements, though it is probable that the
travelling wave electro-optic modulator has a higher bandwidth.   However, the demand for amplitude modulators is as
yet quite modest, due to the slow growth of optical
communications and laser display techniques.

More recently acousto-optics has been applied with
success to the Q-switching of continuously pumped Nd:YAG
lasers.   A fused silica or glass device within the laser
diffracts light from the beam and reduces the laser Q.
When the acoustic power is cut off, the losses are very low
and the Q of the laser rises, allowing a giant pulse of
radiation to be produced.   The switching time (τ) is

typically 250 ns.   In this particular case the advantage lies within the low insertion loss of the device in the un-energised state, rather than a high efficiency of diffraction.   For this reason, readily available glasses of high optical quality are preferable to the more efficient but optically imperfect materials.   Development of materials for this device is concentrated on the further reduction of optical absorption to extremely low levels to cope with the very high optical power densities in high powered lasers.

These acousto-optic Q switches are now commercially available and are finding application in laser micro-machining, resistor trimming, and similar tasks in the electronics industry.

Beam deflection is achieved by altering the frequency (f) of the input power to the transducer in a device such as that shown in Fig.2, causing a change in the angle θ.   The actual magnitude of the change ($\Delta\theta$) is not so important as the number of resolvable beams (N), which is the ratio of the angular change ($\Delta\theta$) to the angle of diffraction (λ/d) appropriate to the aperture (d), i.e.

$$N = \frac{\Delta\theta}{(\lambda/d)} .$$

As the frequency changes the Bragg condition will not be obeyed exactly, but it can be shown that if

$$L = \frac{nv}{f\,\Delta\theta} = \frac{nvd}{fN\lambda}$$

it is possible to sweep over the angle ($\Delta\theta$) by a change of frequency

$$\Delta f = \frac{nv^2}{\lambda\,f\,L} .$$

From these expressions it can be shown that

$$N = \tau \, \Delta f \, .$$

It can be seen that for a large number of resolvable beam
positions a large switching time and large bandwidths are
required.   In situations where the deflection is continuous,
the value of τ is not always important;   for random switching,
e.g. in an optical computer store, it could be important to
have τ very small.   Unfortunately a large N and small τ are
incompatible for a given bandwidth.

The fractional bandwidth $\Delta f/f$ of these devices is
usually set by the transducers and the acoustic mismatch
between the transducers and the acousto-optic material.   In
practice it is possible to achieve useful fractional bandwidths of up to 50% in most solid materials using lead
zirconate-titanate or lithium niobate transducers.   The
centre frequency (f) is set by the required bandwidth subject
to sufficiently low absorption of the material and adequate
power handling capability by the transducers.   Taking
$\Delta f = 0.5f$ as a practical limit, it can be seen that the
length L of such a device should be

$$L = \frac{2nv^2}{\lambda f^2}$$

Referring back to diffracting efficiency, it is
possible to show that for devices of this length the basis
of comparison for materials is M_4 when acoustic power density
is the limiting factor, and M_3 when total power is the
limiting factor.   In some circumstances it is possible to
use non-circular laser beams, by means of cylindrical optics,
and achieve a higher performance with less power.   In these
cases another figure of merit $M_1 = (n^7 p^2/\rho v)$ is appropriate.

to steer the acoustic beam within the material as the
frequency is changed, in order to maintain the Bragg
condition.   Such devices do increase the resolution, but
the complication involved in the manufacture of the transducer assembly, and the problems associated with the wide
band electrical matching to the electronic system are
formidable.

At frequencies above 300 MHz monolithic piezoelectric
transducers are too thin to be practicable, and thin film
ZnO or CdS transducers made by vacuum deposition are usually
employed.   The power handling capability of these devices
is low as they are susceptible to dielectric breakdown and
this limits the power density P^*.   To increase the
efficiency of devices at these frequencies, the transducers
can be put down onto a cylindrical surface so that the
acoustic waves can be focussed into the narrow region
occupied by the laser beam.   This technique has been used
to advantage for modulators and switches with bandwidths of
several hundred MHz using fused silica.

There are other applications of acousto-optic interactions which should be mentioned for completeness.
Acousto-optic diffraction can be used for signal processing,
which involves a combination of modulation and deflection,
to enable information to be extracted optically from complex
signals applied to the transducer.   The technique is also
used in optically accessed acoustic delay lines.

In the last year there has been considerable activity
in the interaction of light with surface acoustic waves in
various configurations, some of them using light travelling
in optical waveguides.   It could well be that this type of
acousto-optic device will have distinct advantages in some
applications.

Beam deflection devices for such applications as optical radars, television laser displays and flying spot microscopes have been made using water as the active medium. Typically they work at about 20 MHz with a bandwidth of ~ 3 MHz.　This small fractional bandwidth is due to the difficulty of coupling acoustic waves into water which has a low acoustic impedance.　With an aperture of ~ 5cms the transit time (τ) is $\simeq$ 30 μs.　This gives a maximum theoretical resolution N $\simeq$ 100.　The optimum length L is $\simeq$ 6cms. Obviously, water is one of the few materials available in such sizes.　Diffraction efficiency of about 90% into the deflected beam is achieved with about one watt of acoustic power.　It can be seen that acousto-optic diffraction cells based on water are efficient, cheap, but limited in resolving power and rather slow.

A more recent solid state device uses $PbMoO_4$ with a centre frequency of about 100 MHz and a bandwidth of 50 MHz. With an aperture of 4cm the switching time is 10 μs and the number of resolvable spots 500.　The optimum length in this case is about 1.5cm.　Crystals of sufficient size are now being grown in acceptable quality.　Because of the crystal orientation required in this device, it is necessary to use a non-circular beam which also gives some economy in acoustic power.　In the other dimension the beam is typically 5mm wide.　In such a device almost 100% diffraction can be obtained for about one watt of acoustic power.　Beam deflectors of this specification have recently appeared on the commercial laser market.

Attempts have been made to circumvent the limitations on N outlined above by the use of special techniques. Using multiple transducers in a stepped array it is possible

V.   CONCLUSIONS

Some of the best materials for acousto-optic devices used in the visible and near infrared are oxides.   The most promising are the double oxides, one being a transition metal oxide and the other a heavy metal oxide.   In general they are dense, soft, and melt in the range 800 to 1300°C. There is every reason to hope that more of these materials will be discovered.

A most welcome development would be the emergence of acousto-optic materials with fewer growth problems that can be produced repeatably in large sizes with the optic quality of glasses.   At the present time the optical homogeneity of the more useful crystals is not good enough for their full device potential to be realised.

GENERAL REFERENCES

Adler, R. (1967) I.E.E.E. Spectrum, 42-54.

Dixon, R.W. (1967) J. App. Phys. 38, 5149-5153.

Dixon, R.W. and Cohen, M.G. (1966) Appl. Phys. Letters 8, 205-206.

Gordon, E.I. (1966) Proc. I.E.E.E. 54, 1391-1401.

Klein, W.R. and Cook, B.D. (1967) I.E.E.E. Trans. Son. Ultrason. SU-14, 123-134.

Korpel, A., Adler, R., Desmares, P., and Watson, W. (1966) Proc. I.E.E.E. 54, 1429-1437.

Nye, J.F. (1960) "Physical Properties of Crystals", Clarendon Press, Oxford.

Ohmachi, Y. and Uchida, N. (1971) App. Phys. 42, 521-524.

Pinnow, D.A. and Dixon, R.W. (1968) Appl. Phys. Letters 13, 156-158.

Pinnow, D.A., Van Uitert, L.G., Warner, A.W. and Bonner, W.A.
   (1969) Appl. Phys. Letters 15, 83-86.
Uchida, N. and Ohmachi, Y. (1970) Jap. Appl. Phys. 9, 155-156.
Wemple, S.H. and DiDomenico, M. (1970) Phys. Rev. B1, 193-202.

OXIDES IN NON-LINEAR OPTICS

K.F. HULME

Royal Radar Establishment,
Malvern, Worcestershire

I.   THE PHENOMENA OF NON-LINEAR OPTICS
    A.   INTRODUCTION
    B.   LINEAR OPTICS
    C.   OPTICAL FREQUENCY MIXING
    D.   THE ELECTRO-OPTIC EFFECT

II.   CRYSTAL PROPERTIES REQUIRED

III. CRYSTALS CURRENTLY USED IN NON-LINEAR OPTICS

IV.   THE PREPARATION OF OXIDE CRYSTALS FOR
    NON-LINEAR OPTICS

V.   APPLICATIONS OF NON-LINEAR OPTICS
    A.   OPTICAL FREQUENCY MIXING DEVICES
    B.   ELECTRO-OPTIC DEVICES

I.   THE PHENOMENA OF NON-LINEAR OPTICS

A.   INTRODUCTION

A restricted view of non-linear optics is taken, but
the practical applications for which oxides have importance
are included.   Attention is confined to two aspects:
optical frequency mixing, and the electro-optic effect.
Optical frequency mixing is concerned with the generation of

sum and difference optical frequencies when optical radiation impinges on crystals.   The electro-optic effect is the change of refractive index produced in a crystal by an electric field.   The optical part of the spectrum is considered here to include wavelengths in the range 0.2 - 20μm. These two aspects of non-linear optics arise from the same underlying non-linear relationship.

The non-linear relationship is that between $\underline{P}$, the polarisation (or electric dipole moment per unit volume) of a medium, and $\underline{E}$, the applied electric field.   The optical frequency components of $\underline{E}$ and $\underline{P}$ are related by the series expansion

$$\underline{P} = \alpha\underline{E} + \beta\underline{E}^2 + \gamma\underline{E}^3 + \ldots \qquad \ldots(1)$$

For a centrosymmetrical medium $\beta \equiv 0$, and the most important non-linear term vanishes.   Since gases, normal liquids, glasses and fine-grained polycrystalline material are centrosymmetrical, non-linear optics is concerned almost exclusively with the small number of crystals that lack a centre of symmetry, i.e. are acentric.

B.   LINEAR OPTICS

Linear optics concerns effects arising from the $\underline{E}$ term of Eq.(1).   It follows from Maxwell's equations that the linear optical properties of the simplest isotropic crystal (namely the refraction of waves at its surface and their phase velocity inside) are described by the refractive index, $\underline{n}$, which is related to α by

$$\underline{n} = \sqrt{(1 + 4\pi\alpha)} \qquad \ldots(2)$$

The interior phase velocity is $\underline{n}$ times smaller than that in

vacuum.   Absorption can be ignored, but the fact that
crystals can be optically anisotropic must be considered.
Uniaxial crystals are the simplest anisotropic case which,
for a given propagation direction and optical frequency,
exhibit two possible waves called 'ordinary' and 'extraordinary'.   These waves have electric fields orthogonal to
one another and to their common propagation direction, thus
corresponding to orthogonal linear polarisations.   The
directions of the electric fields for the two waves constitute the so-called vibration directions for that propagation
direction (see Fig.1).   The two waves have different phase
velocities, i.e. different refractive indices.   For a given
frequency the ordinary refractive index, $n_{ord.}$, does not
depend on direction whereas the extraordinary index, $n_{ext.}$,
is orientation dependent.   There is thus some freedom of
choice of phase velocity in anisotropic crystals but both
refractive indices depend on frequency.

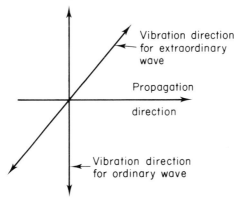

Vibration direction
for extraordinary
wave

Propagation

direction

Vibration direction
for ordinary wave

*Fig.1.   The relationship in a uniaxial crystal between
the light propagation direction and the vibration
directions of the ordinary and extraordinary waves;
the three directions are mutually orthogonal.*

C. OPTICAL FREQUENCY MIXING

These effects arise from the $\beta \underline{E}^2$ term of Eq.(1) by putting

$$\underline{E} = \underline{E}_1\cos(\omega_1\underline{t} - \underline{k}_1\underline{x}) + \underline{E}_2\cos(\omega_2\underline{t} - \underline{k}_2\underline{x}) \qquad ...(3)$$

thus considering two monochromatic waves propagating along the $\underline{x}$ direction with angular frequencies ω_1 and ω_2 and wave vectors $\underline{k}_1 = \omega_1\underline{n}_1/\underline{c}$ and $\underline{k}_2 = \omega_2\underline{n}_2/\underline{c}$ (where $\underline{c}$ is the velocity of light in vacuum, and $\underline{n}_1$ and $\underline{n}_2$ are the refractive indices at ω_1 and ω_2). From the term $\beta \underline{E}^2$ polarisation waves are obtained at frequencies 0, $2\omega_1$, $2\omega_2$, $(\omega_1 + \omega_2)$ and $(\omega_1 - \omega_2)$. Consider the term at frequency $(\omega_1 + \omega_2)$

$$\beta\underline{E}_1\underline{E}_2 \cos\{(\omega_1 + \omega_2)\underline{t} - (\underline{k}_1 + \underline{k}_2)\underline{x}\} \qquad ...(4)$$

This sum-frequency polarisation wave will radiate an electromagnetic wave. In order to evaluate the output field at $\underline{x} = \underline{L}$, for a slab extending from $\underline{x} = \underline{0}$ to $\underline{x} = \underline{L}$ the contributions to the electromagnetic field at $\underline{x} = \underline{L}$ must be considered for all the infinitesimally thin slices which constitute the slab. Clearly the output field will be maximised if all contributions add constructively. Now the propagation from any slice to the surface $\underline{x} = \underline{L}$ proceeds with the phase velocity appropriate to the frequency $(\omega_1 + \omega_2)$. Write $\omega_3 \equiv \omega_1 + \omega_2$; $\underline{n}_3$ then signifies the refractive index at ω_3. The time delay in propagating to $\underline{x} = \underline{L}$ from the slice between $\underline{x}'$ and $\underline{x}' + \delta\underline{x}'$ is $(\underline{L} - \underline{x}')/(\underline{c}/\underline{n}_3) \equiv (\underline{L} - \underline{x}')(\underline{k}_3/\omega_3)$, where $\underline{k}_3 = \omega_3\underline{n}_3/\underline{c}$ is the wave vector for the ω_3 wave; using Eq.(4), the contribution from this slice is, therefore, proportional to

$$\beta\underline{E}_1\underline{E}_2 \cos\{\omega_3(\underline{t} - (\underline{L} - \underline{x}')(\underline{k}_3/\omega_3)) - (\underline{k}_1 + \underline{k}_2)\underline{x}'\}\delta\underline{x}' \qquad ...(5)$$

If the condition

$$\underline{k}_3 = \underline{k}_1 + \underline{k}_2 \qquad \ldots(6)$$

can be satisfied the contribution of all slices add constructively, because in this case the argument of the cosine
in Eq.(5) loses all dependence on $\underline{x}$.   Condition Eq.(6) is
known as the phase-matching condition.   If this condition
is satisfied, the optimum sum-frequency output field is
radiated.   Using $\underline{k} = \omega_1 \underline{n}_1 / \underline{c}$, etc., Eq.(6) can be rewritten
as

$$\omega_3 \underline{n}_3 = \omega_1 \underline{n}_1 + \omega_2 \underline{n}_2 \qquad \ldots(7)$$

If $\underline{n}$ did not depend on ω, Eq.(7) would be satisfied identi-
cally; but $\underline{n}$ does depend on ω, so that a stratagem must be
used to satisfy Eq.(7) and obtain efficient sum-frequency
generation.   The stratagem is to offset the change of $\underline{n}$
with ω by appropriate change of polarisation (ordinary to
extraordinary) and a choice of propagation direction.   Even
this will only work if the crystal has a rather marked
optical anisotropy.   Thus, the phase-matching condition
might be satisfied in a sufficiently anisotropic crystal by
making wave 1 and 2 ordinary, and taking the output wave 3
as extraordinary.   For maximum output by phase-matching to
be achieved the condition

$$\omega_3 \underline{n}_{3\text{ext.}} = \omega_1 \underline{n}_{1\text{ord.}} + \omega_2 \underline{n}_{2\text{ord.}} \qquad \ldots(8)$$

must be satisfied so a precise knowledge of the ordinary and
extraordinary refractive indices of the crystal being
considered, and any variations with frequency, are required
before the possibility of phase-matching can be predicted.

Several important points arise from these considerations. Firstly, difference-frequency generation is similar in principle to sum-frequency generation. Secondly, second-harmonic generation can be regarded as a special case of sum-frequency generation (in which $\omega_1 = \omega_2$). Thirdly, large output powers are favoured by large values of β and by large input powers; in fact, only when ω_1 and/or ω_2 is provided by a laser are output powers large enough to be readily observable. Fourthly, the term $\gamma \underline{E}^3$ in Eq.(1) also produces frequency mixing but in practice the effects stemming from $\beta \underline{E}^2$ are of far greater importance.

More detailed treatments of optical frequency mixing are given by Yariv, (1968); Minck et al. (1966); Gibson, (1968).

D.   THE ELECTRO-OPTIC EFFECT

Returning to Eq.(1), the condition

$$\underline{E} = \underline{E}_0 + \underline{E}_1 \sin (\omega_1 \underline{t} - \underline{k}_1 \underline{x}) \qquad \ldots(9)$$

corresponds to applying a DC field to a crystal through which monochromatic light is propagating. From the term $\beta \underline{E}^2$, a contribution to the polarisation at frequency ω_1 with magnitude $2\beta \underline{E}_0 \underline{E}_1 \sin (\omega_1 \underline{t} - \underline{k}_1 \underline{x})$ is obtained. This can be added to the contribution $\alpha \underline{E}_1 \sin (\omega_1 \underline{t} - \underline{k}_1 \underline{x})$ arising from $\alpha \underline{E}$. It is seen that the application of $\underline{E}_0$ has had the effect of changing α to $(\alpha + 2\beta \underline{E}_0)$; correspondingly, from Eq.(2), the refractive index changes by an amount

$$\delta \underline{n} \simeq 4\pi \beta \underline{E}_0 / \underline{n} \qquad \ldots(10)$$

proportional to $\underline{E}_0$. This is the linear electro-optic effect, or Pockels' effect, known long before the advent of the laser.

For a uniaxial crystal, both $n_{ord.}$ and $n_{ext.}$ can, in general, be changed by E_0. The changes depend on field magnitude and direction and on propagation direction. A convenient way to specify the electro-optic effect for a given configuration in a given crystal is to specify the half-wave voltage, which is defined as follows.

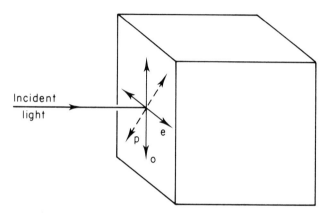

Incident
light

Fig.2. A unit cube of a uniaxial electro-optic material (o = ordinary and e = extraordinary vibration directions; p = plane of polarisation equally inclined to the two vibration directions).

Consider the cube of Fig.2. Light propagates normal to one face. The input light is plane polarised at 45° to the ordinary and extraordinary vibration directions. Consequently, the ordinary and extraordinary waves are in phase at the input surface and have equal magnitude. Consider first the zero DC field situation. At the output surface the two waves will not generally be in phase because the phase difference, in radians $(2\pi/\lambda(n_{ord.} - n_{ext.})$, will not generally be an integral multiple of 2π. However, suppose that, by a small adjustment of temperature, the refractive indices are altered so that $(2\pi/\lambda)(\underline{n}_{ord.} - \underline{n}_{ext.})$ = $2N\pi$, where N is an integer, the output polarisation would then correspond to the input.   If an electric field is applied between the top and bottom faces of the cube, the half-wave voltage (for that crystal with that configuration of field and propagation directions) is defined as the voltage needed to displace the phases of the two waves by π radians at the output surface.   If the phases are displaced by π this changes the plane of polarisation from its original position to the plane at right angles (see Fig.3).

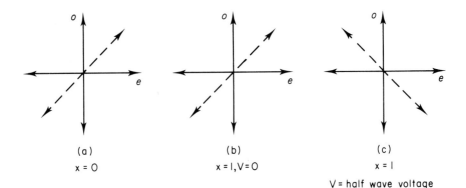

(a)   (b)   (c)

x = 0   x = l,V = 0   x = l

V = half wave voltage

Fig.3. States of polarisation of light (broken lines) under conditions applicable to Fig.2.
a) Input face.   b) Output face, zero applied voltage, voltage-independent phase difference of 2Nπ.   c) As b), but with applied voltage equal to the half-wave voltage.

A high value of β leads to a small half-wave voltage.   If the crystal is followed by an analyser, crossed relative to the input polarisation, the application of the half-wave voltage will switch the light transmitted by the analyser from off to on.

With a cube geometry electrodes can be applied to consider fields perpendicular to the light propagation direction or parallel to it.   Electro-optic effects exist when electric field and propagation directions have more general mutual relations, and the cube geometry needs to be modified to consider them;   these cases are not currently important.

So far DC applied fields have been considered. Alternating fields with frequencies up to the GHz range also produce electro-optic effects.   In this way amplitude modulation or phase-modulation of light at the impressed frequency can be obtained.

If the term $\beta \underline{E}^2$ in Eq.(1) is zero (centrosymmetric medium), the effects from $\gamma \underline{E}^3$ will perhaps be observable. This term will give a quadratic electro-optic effect, or Kerr effect, in which refractive index and phase changes become proportional to the squares of applied voltages. The effects are of limited practical importance.

In a real crystal the values of β in Eq.(1) appropriate for optical frequency mixing ($\beta_{o.f.m.}$) and for the electro-optic effect ($\beta_{e.o.}$) can differ greatly (just as the radio-frequency and optical frequency dielectric constants can differ greatly).

The electro-optic effect has been reviewed recently by Kaminow and Twiner (1966);   Hulme (1969).

II.   CRYSTAL PROPERTIES REQUIRED

Optical frequency mixing crystals should be:

(i)   acentric, otherwise $\beta_{o.f.m.} \equiv 0$ and the significant
       effects vanish,

(ii)   optically transparent for the spectral region of
       interest,

(iii) markedly optically anisotropic in order to provide
       ample phase-matching opportunities; in a specific
       phase-matching situation (e.g. generation of second
       harmonic of a neodymium laser) there is great
       advantage in having very nearly the exact crystal
       anisotropy so that small adjustments can be made by
       varying the crystal temperature,

(iv)   markedly non-linear, i.e. a large $\beta_{o.f.m.}$; there is
       an empirical rule stating that $\beta_{o.f.m.}$ is roughly
       proportional to $(\underline{n}^2 - 1)^3$ which suggests that large
       values of $\underline{n}$ are desirable,

(v)   available in interferometric optical quality in
       centimetric sizes, although smaller crystals may have
       enough other advantages to be important; refractive
       index changes of $\sim 10^{-5}$ are important, because they
       destroy phase-matching,

(vi)   durable in normal ambient situations and under high
       laser powers.

Electro-optic crystals should be:

(i)   acentric; this is not obligatory but strongly pre-
       ferred, since effects from $\beta_{e.o.} \underline{E}^2$ usually predominate
       over those from $\gamma \underline{E}^3$ if $\beta_{e.o.} \neq 0$,

(ii) optically transparent for the spectral region of
      interest,

(iii) optically isotropic; again this is preferred but is
      not essential.   The preference exists because the
      smaller the crystal anisotropy, the less the problem
      in designing devices with large angular aperture;
      also, when using anisotropic crystals, it is often
      necessary either to control the crystal temperature
      or to use a temperature-compensating pair of crystal
      specimens to avoid undesirable phase changes caused
      by refractive index alterations as the temperature
      changes,

(iv) markedly non-linear, i.e. large $\beta_{e.o.}$ and thus small
      half-wave voltage; crystals with a large radio-
      frequency dielectric constant   have a large $\beta_{e.o.}$,

(v)   available in interferometric quality in centimetric
      sizes as for optical frequency mixing crystals,

(vi) durable as for optical frequency mixing crystals.

The notable feature of the two lists of requirements
is the number of points in common, i.e. those labelled (i),
(ii), (v) and (vi).   Even the points of difference (iii)
and (iv) do not prevent a crystal from being important for
both aspects.   Thus, the two aspects of non-linear optics
have a unity in their working substances as well as in their
underlying physics.

There is no logical chain of argument leading from
these requirements to a list of chemical formulae, which
would give crystals to satisfy the requirements in the
fullest possible way.   A few links of the chain sometimes
exist, for instance ferroelectrics are acentric, have high

values of ϵ near Curie points and useful electro-optic properties.   For durability, oxygen-octahedral ferroelectrics
(e.g. niobates) would be chosen.   Inspection of a list of
favoured electro-optic crystals lends some support to such a
line of argument as niobates are well represented;   however,
they are not exclusively oxygen-octahedral ferroelectrics.
Furthermore, oxygen octahedral ferroelectrics have the
disadvantage that they are often difficult to grow in the
required size and quality so even if a chemical formula can
be predicted, the quality of its crystal form cannot.
Water-soluble crystals can sometimes be grown from aqueous
solution to great size and perfection, but this is not a
general rule; $KBrO_3$ is a well-known exception.   Growth
from the melt is also capable of producing crystals for
optics (e.g. CaF_2) but the difficulty of producing really
large optical quality crystals of ferroelectrics, such as
barium sodium niobate, is generally acknowledged.

    Some generalisations about the spectral transmission
range of substances can, however, be made.   Oxides, even
complex oxides, have energy gaps which are usually large
enough not to interfere with transmission in the visible.
For sulphides, absorption often sets in (proceeding to
shorter wavelengths) somewhere in the visible, whilst
selenides are often opaque in the same region.   The infra-
red transmission of crystals is limited by the existence of
lattice vibrations;   if hydrogen is present, the crystal may
well be opaque for wavelengths longer than about 1 m.
However, if oxygen is the lightest atom present, the transmission does not decrease appreciably until the wavelength
exceeds $\sim$ 5μ.   If sulphur is the lightest atom present,
the transmission limit can be beyond 15μm.   Oxides,

therefore, play a role in the spectral range 0.4 - 4µm.

III. CRYSTALS CURRENTLY USED IN NON-LINEAR OPTICS

Table I gives a representative list of currently important crystals.   It is not an exhaustive list, but no crystal of known importance is excluded.   More comprehensive lists have been compiled by Bechmann (1969); Bechmann and Kurtz (1969).

The crystals recorded in Table I are those available in a usable form.   Substances known to have desirable properties, but impossible to grow as useful single crystals are excluded.

It is premature to draw firm conclusions about the relative importance of oxides as the discovery of new materials is still substantially by empirical methods.   If the phosphates, arsenates and iodates are excluded as being salts of oxyacids rather than oxides in the conventional sense, the oxides left in Table I are niobates or tantalates. These crystals are examples of the oxygen-octahedral ferro-electrics already mentioned; they are very non-linear and for $LiNbO_3$ and $Ba_2NaNb_5O_{15}$ phase-matching conditions are particularly favourable for second-harmonic generation from the 1.06µm Nd^{3+} laser to the green 0.53µm.   The future of these materials is very dependent upon the expense and yield of the crystal growth methods used.   The simple double oxide niobate $LiNbO_3$ is easier to grow than the more complex ternary oxides $Ba_2NaNb_5O_{15}$ and $K(Ta,Nb)O_3$; the difference is related to the phase diagrams.   Lithium niobate exists over a relatively small range of composition ($\sim$ 5 mol.% solid solubility range 50°C below max. m.pt.), whereas for barium sodium niobate the compositional range is large

TABLE I

Name of Substance	Potassium dihydrogen phosphate	Potassium dideuterium phosphate	Caesium dideuterium arsenate	Lithium niobate	Barium Sodium niobate	Potassium tantalo-niobate	Lithium iodate	Mercury cadmium thiocyanate	Proustite	Tellurium
Nominal Chemical Formula	(KH_2PO_4)	(KD_2PO_4)	(CsD_2AsO_4)	$(LiNbO_3)$	$(Ba_2NaNb_5O_{15})$	$(K(Ta_x Nb_{1-x})O_3)$	$(LiIO_3)$	$(HgCd(CNS)_4)$	(Ag_3AsS_3)	(Te)
Pt.Gp.Symmetry	$\bar{4}2\,m$	$\bar{4}2\,m$	$\bar{4}2\,m$	3m	4mm	> m3m (Curie-point)	6	$\bar{4}$	3m	32
Curie-point (°C) (if ferroelectric)	-150	-60	-61	1200	560	0 for x = 0.65	-	-	-	-
Useful for - o.f.m.	Yes	Yes	Yes	Yes	Yes	No	Yes	Yes	Yes	Yes
e.o.d.	Yes	Yes	No	Yes	-	-	No	-	No	No
Transmission range (µm)	0.2-1.0	0.2-1.5	0.2-1.5	0.4-4.0	0.4-4.0	0.4-4.0	0.3-5.0	0.4-2.5	0.6-13.0	4.5-20.0
Non-linear coeff. for o.f.m. use (mksu x 10^{24})	5	5	5	30, 60	140, 170	-	60	20, 80	150, 250	8,000
Electro-optic half-wave voltage at ~ 0.6µm (kV)	8	4	2	2.7	1.3	0.5*	40	-	9	opaque in visible
field along:-	{001}	{001}	{001}	{0001}	{001}	{001}	{0001}		{0001}	
light along:-	{001}	{001}	{001}	{101̄0}	{100}	{100}	{101̄0}		{101̄0}	
Growth method	aqueous solution			melt			aqu.soln.	solution	melt	melt

* for first half wave retardation; 1cm cube specimen

($\sim$ 20 mol.% solid solubility on the $BaNb_2O_6$ - $NaNbO_3$ pseudo-
binary 50°C below max. m.pt.) and for $K(Ta,Nb)O_3$ complete
solid solubility extends across the phase diagram.   The
melt-grown niobates and tantalates are in general less easy
to produce than crystals grown from aqueous solution which
could lead to more emphasis on water-soluble materials.

One side effect of the successful search for improved
non-linear optical crystals has been to provide new crystals
for older fields.   Thus, lithium niobate has established
itself as a useful piezoelectric crystal (piezoelectrics
also have to be acentric).

IV.   THE PREPARATION OF OXIDE CRYSTALS FOR
       NON-LINEAR OPTICS

Attention is restricted to the niobates and tantalates.
The crystals are invariably grown from the melt.   In the
case of $LiNbO_3$ (m.pt. $\sim$ 1250°C) and $Ba_2NaNb_5O_{15}$ (m.pt.
1450°C) the Czochralski method is used.   The melt is contained in a platinum crucible, heated by r.f. induction.
The melt can be made directly from the alkali or alkali
earth metal carbonates (or nitrates) and niobium pentoxide;
alternatively, it may be composed of portions selected from
a previously-solidified melt.   The first method of forming
the melt has the disadvantage that it is difficult to arrive
at a nominated melt composition.   The second method has the
advantage that previous crystallisations should have taken
the melt composition towards the maximum melting point,
which does not occur exactly at the composition given by the
nominal chemical formula;   thus, the composition of the
growing crystal should be less susceptible to fluctuations
in stirring and growth rate, and better optical quality

should be obtained in the grown material.   Repeated crystallisation is likely to be more of a necessity for $Ba_2NaNb_5O_{15}$
growth than for $LiNbO_3$, because the liquidus curvature at
the maximum melting point is smaller and the solid solution
range is larger for the former substance than it is for the
latter.

K(Ta_xNb_{1-x})O_3 crystals are also grown from the melt
(Wilcox and Fulmer, 1966).   The melt is made from K_2CO_3,
Ta_2O_5 and Nb_2O_5;   a platinum crucible is again used, often
heated by a resistance furnace, and for $x = 0.65$ a temperature near $1100^{\circ}C$ is needed.   The linear growth velocity is
very low, since the process is essentially one of growth
from solution, and the solid has a ratio of Ta to Nb
different from that in the melt.   The use of recrystallised
melts is thus not feasible.   The ratio of Ta to Nb in the
grown crystals is very susceptible to fluctuations in growth
conditions.

For $LiNbO_3$ and $Ba_2NaNb_5O_{15}$ (but not usually for
K(Ta_xNb_{1-x})O_3), one is using the material below its Curie
point.   The possibility of multidomain material has therefore to be taken into account.   The applications invariably
demand single domain material;   for $LiNbO_3$ melt-grown
material is often single domain;   but this is not the case
for $Ba_2NaNb_5O_{15}$.   Multidomain material can be converted to
single domain by applying an electric field while holding
the crystal near its Curie point;   the process is known as
poling.   When niobates and tantalates with elevated Curie
points are poled, appreciable current often flows through
the crystal.   The resistance also changes as poling proceeds
and it is known that hydrogen migrates through the crystal
during the process.   Clearly poling in these materials has

both chemical and physical aspects, a possibility not previously envisaged in the poling of ferroelectrics. From a practical viewpoint poling is not a serious difficulty but it does add another process to the production chain for usable material. In the case of $Ba_2NaNb_5O_{15}$ there is yet another process to perform, namely, detwinning. This necessity arises from the transition near $200^\circ C$ in which the structure changes from tetragonal to orthorhombic. Detwinning is accomplished by applying an appropriately directed stress to the crystal as it cools through the transition temperature. After poling and detwinning (if needed) have been completed, orientated specimens may be cut and polished without difficulty.

Preferred methods of preparing $LiNbO_3$ and $Ba_2NaNb_5O_{15}$ have been recently described respectively by Byer et al. (1970) and by Barraclough et al. (1970).

V.   APPLICATIONS OF NON-LINEAR OPTICS

A.   OPTICAL FREQUENCY MIXING DEVICES

1. Second harmonic generation

This is particularly important for doubling 1.06 to $0.53\mu m$ when the following remarks apply; barium sodium niobate is needed for efficient continuous conversion (peak power low); for high pulse repetition rates (intermediate peak powers) lithium iodate may be preferred; for high peak powers the arsenates or phosphates may be needed, since they have higher laser damage thresholds; lithium niobate is not used, except at low mean power levels because the green light creates refractive index inhomogeneities which spoil phase-matching.

2. Optical parametric oscillators

This topic has been reviewed by Harris (1969). Oscil-
lators of this type are tunable sources based on a pumping
laser and a non-linear (o.f.m.) crystal. The crystal is in
a cavity (Fig.4) and when the pump laser is on, optical
oscillations at two frequencies, ω_1 and ω_2, can build up
inside the cavity; their frequencies add up to the pump
frequency ω_3. The oscillation frequencies, ω_1 and ω_2, can
be altered by adjusting the phase-matching conditions, e.g.
by twisting the crystal inside the cavity. A commercial
oscillator tunable from 0.6 to 3μm uses lithium niobate and
lithium iodate.

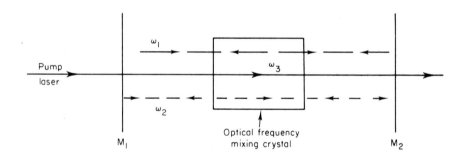

*Fig.4. Optical parametric oscillator $\omega_3 = \omega_1 + \omega_2$;
M_1 and M_2 are mirrors assumed transparent at ω_3 and
highly reflecting at ω_1 and ω_2.*

3. Up-conversion

Warner (1971) has surveyed up-conversion of infra-red
radiation to the visible which uses a sum frequency process

(Fig.5).   The infra-red frequency is added to that of a
visible or near visible laser.   Images can be up-converted.
Up-conversion of radiation from the wavelength region near
10μm may turn out to be the process of the greatest importance;   in that case, oxides could not be used (they are
usually opaque at 10 microns) and proustite is the only
practical crystal known at present.

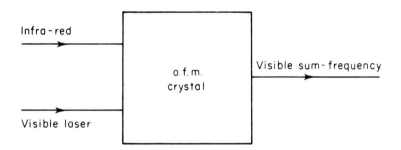

*Fig.5.   Up-conversion of an infra-red
signal to a visible signal.*

B.   ELECTRO-OPTIC DEVICES

1.   Laser Q-switches

Here the electro-optic crystal performs a voltage-
controlled switching action inside a laser cavity.   The
most favoured material for devices commercially available
at present is potassium dideuterium phosphate, but improvements in lithium niobate quality are enabling it to compete
more effectively at infra-red wavelengths, where lithium
niobate does not suffer refractive index changes under
illumination.

K. F. Hulme

2. Modulators in light deflectors

The electro-optic switching action is used to change the polarisation of a laser beam; the beam then falls on a calcite prism which deflects it to one of two directions, depending on its polarisation. Arrays of such elements in optical series are of interest in holographic computer stores. At present potassium dideuterium phosphate is used as the electro-optic material, but potassium tantalo-niobate is sometimes considered as a possible replacement.

3. Modulators for optical beams

These have been reviewed by Chen (1970) and are of interest for optical communication. Lithium niobate, or tantalate, or potassium dihydrogen phosphate or the dideuterium isomorph have been used in commercially available devices.

REFERENCES

Barraclough, K.G., Harris, I.R., Cockayne, B., Plant, J.G. and Vere, A.W. (1970) J. Mat. Sci. 5, 389.

Bechmann, R. (1969) Landolt Börnstein Tables New Series Group III Vol.2 Section 4, p.126, Springer Verlag, Berlin.

Bechmann, R. and Kurtz, S.K. (1969) Landolt Börnstein Tables New Series Group III Vol.2 Section 5, p.167, Springer Verlag, Berlin.

Byer, R.L., Young, J.F. and Fiegelson, R.S. (1970) J. Appl. Phys. 41, 2320.

Chen, F.S. (1970) Proc. I.E.E.E. 58, 1440.

Gibson, A.F. (1968) Science Progress 56, 479.

Harris, S.E. (1969) Proc. I.E.E.E. 57, 2096.

Hulme, K.F. (1969) Electronic Components 10, 69.

Kaminow, I. and Twiner, E.H. (1966) Applied Optics 5, 1612.

Minck, R.W., Terhune, R.W. and Wang, C.C. (1966) Proc.
I.E.E.E. 54, 1357.

Warner, J.W. (1971) Opto-electronics 3, 37.

Wilcox, W.R. and Fullmer, L.D. (1966) J. Amer. Ceramic Soc.
49, 415.

Yariv, A. (1968) "Quantum Electronics" John Wiley, New York.

OXIDES FOR DELAY LINE APPLICATIONS

M.F. LEWIS

The General Electric Company Limited,
Wembley, Middlesex

I.   INTRODUCTION

II.   DELAY LINES EMPLOYING PIEZOELECTRIC TRANSDUCERS
- A. BULK WAVE TRANSDUCER MATERIALS
- B. SURFACE WAVE TRANSDUCER MATERIALS
- C. THE ROLE OF OXIDES AS TRANSDUCER MATERIALS
- D. BULK WAVE DELAY MEDIA
- E. THE ROLE OF OXIDES AS DELAY MEDIA

III. MAGNETOELASTIC DELAY LINES

IV. SOME APPLICATIONS OF HIGH FREQUENCY DELAY LINES

I.   INTRODUCTION

Most solid state ultrasonic delay lines employ oxide materials as transducers and delay media. These range from fused silica delay lines with quartz or $BaTiO_3$-type transducers at MHz frequencies, to sapphire delay lines with ZnO transducers at microwave frequencies. The main reason for the predominance of the oxides in such devices is the availability of high-quality low-cost materials with large physical dimensions. A second factor is the stability of the oxides which, as well as rendering them immune to attack

74

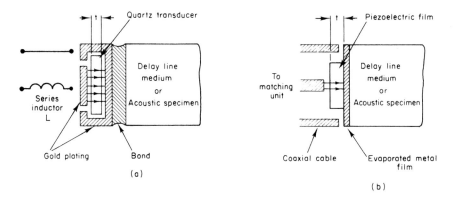

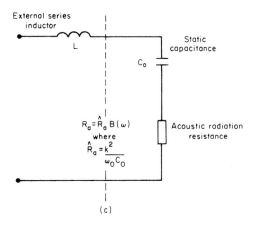

Fig.1. a) Schematic diagram of conventional delay line with bonded transducers, b) with evaporated transducers, c) equivalent circuit for series tuned transducers in a) and b).

by the atmosphere, provides a tightly bound lattice. The
oxides are thus frequently hard dielectrics which are suitable for machining and polishing. They also commonly
exhibit low ultrasonic losses which makes them necessary
components of very high frequency delay lines.

II. DELAY LINES EMPLOYING PIEZOELECTRIC TRANSDUCERS

The basic ultrasonic delay line ordinarily comprises
an input transducer, delay medium, and output transducer,
although it is occasionally possible to work in a reflection
mode with a single transducer. The conventional delay line
employing resonant quartz transducers is shown in Fig.1a and
is useful for operation up to $\sim$ 100 MHz. In Fig.1b is
shown a similar arrangement employing evaporated piezoelectric transducers, usually CdS or ZnO. The advantages
of this latter arrangement are (a) that the electrode and
transducer are vapour deposited directly on to the delay
medium thereby eliminating the bond, and (b) that the transducers can be deposited in very thin layers for resonant
operation at high frequencies, e.g. a 3μ layer resonates at
$\sim$ 1 GHz, while a 0.3μ layer resonates at $\sim$ 10 GHz. Thus
operation up to X-band frequencies is feasible, although
various mechanisms introduce some insertion loss at these
frequencies. A disadvantage of these evaporated transducers is that only longitudinal waves can be excited
efficiently at the highest frequencies.

In Fig.2 is shown a novel interdigital finger transducer (I.D.T.) for the excitation of surface (acoustic)
waves on piezoelectric media (White and Voltmer, 1965).
The electric field configuration of the I.D.T. generates a
complex strain field, but the net surface wave excitation

can be optimised at frequency ω_0 by making the period of the
I.D.T. equal to the wavelength $\lambda_0 = 2\pi c/\omega_0$ where c is the
velocity.   In these circumstances the wavelets of the surface wave generated at each gap interfere constructively.
In contrast to the bulk wave devices of Fig.1 the number of
acoustic sources can be chosen to optimise the efficiency-
bandwidth product of the transducer.   Using conventional
photo-lithographic techniques the finger and gap widths can
each be made as small as 1μ, corresponding to a wavelength
of $\sim$ 4μ, or a frequency of $\sim$ 1 GHz.   With electron beam
fabrication techniques the upper frequency approaches X-band.

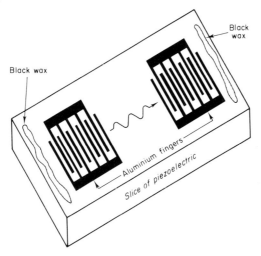

*Fig.2.   Schematic diagram of the surface wave
interdigital transducer (I.D.T.).   The black
wax is included to absorb unwanted waves.*

A.   BULK WAVE TRANSDUCER MATERIALS

   One desirable property of the transducer material is
that it should have a similar acoustic impedance $Z(Z = \rho c$,
where ρ = density, c = sound velocity), to the delay medium

TABLE I

Piezoelectric material	Density ρ (10^3 kg/m³)	Acoustic wave l=longitudinal t=transverse	Acoustic velocity c (m/s)	Acoustic Impedance Z (10^6 kg/m²s)	Square of electromechanical coupling coefficient k^2(%)
Quartz	2.65	l(x-axis)	5700	15.1	0.86
		t(y-axis)	3800	10.0	2.0
Tourmaline	3.1	l(c-axis)	7200	22.3	0.8
CdS	4.8	l(c-axis)	4500	21.6	~2.2
		t(a-axis)	1800	8.6	~3.6
ZnO	5.6	l(c-axis)	6300	35.3	~7.8
		t(a-axis)	2700	15.0	~10.0
LiNbO$_3$	4.7	quasi-l (36° rotated y-cut)	7400	34.8	25
		t(x-axis)	4800	22.6	46

to allow transmission of the ultrasonic energy into the
delay medium without reflections at the interface.   This
being so, the transducer response is strongest at the
fundamental frequency, ω_o and its odd harmonics, and drops
smoothly to zero at the even harmonics, the 3 dB bandwidth
being of order ω_o.   A simplified (series) equivalent circuit of the transducer is shown in Fig.1c in which C_o
represents the static capacitance and R_a the acoustic
radiation resistance (i.e. representing power flow into the
delay medium).   Note that R_a contains the important quantity
k^2, the square of the electromechanical coupling coefficient.
Since the acoustic bandwidth ($\sim \omega_o$) is invariably greater
than the electrical bandwidth of Fig.1c, the latter determines the instantaneous bandwidth.   This can be obtained
from the quality factor, $Q \simeq \omega_o/\Delta\omega \simeq (\omega_o C_o R_a)^{-1}$, giving
$\Delta\omega/\omega_o \sim k^2$.   Values of k^2 and other important quantities
for various piezoelectrics are given in Table I.   Another
important transducer consideration is its ability (or otherwise) to excite selectively longitudinal or shear waves,
since either wave can have the lower losses in the delay
medium.

B.   SURFACE WAVE TRANSDUCER MATERIALS

The acoustic bandwidth of the I.D.T. in Fig.2 can be
deduced by noting that, to a good approximation, each finger
pair generates an acoustic wave independently of the others.
Thus the response at frequency ω can be found by forming the
(vector) sum of the contributions, i.e. an acoustic version
of Cornu's spiral in optics (Jenkins and White, 1957).   The
fractional bandwidth is thus 1/N where N is the number of
finger pairs (corresponding to the resolving power of a

M. F. Lewis

TABLE II

Piezoelectric material	Cut and propagation direction	k^2_{eff} (%)	Optimum instantaneous bandwidth (%)	Velocity c (m/s)	Attenuation (dB/cm at 1GHz)
LiNbO$_3$	Y,$\bar{Z}$	5	25	3480	2
Quartz	Y,X	0.2	5	3160	9
Bi$_{12}$GeO$_{20}$	(110),{$\bar{1}$10}	2.3	16	1600	3
ZnO	X,$\bar{Z}$	1.1	12	2700	10
	Z, arbitrary	0.7	8	2700	10
CdS	X,$\bar{Z}$	0.6	9	1700	50
	Z, arbitrary	small	–	1700	50
PZT	Poled normal to surface, propagation direction arbitrary	4.3	23	2200	20 (at 50 MHz)

All data approximate.   Largely taken from Smith et al. (1969)

diffraction grating comprising N slits).   In contrast, the electrical bandwidth of a series tuned I.D.T. is proportional to N, since the static capacitance $C_o = NC_s$ where C_s is the capacitance per finger pair.   A great advantage of the I.D.T. is that by the correct choice of N, the acoustic and electrical bandwidths can be made equal, which optimises the instantaneous bandwidth (Smith et al. 1969).   By symmetry it can be seen that the I.D.T. is acoustically bidirectional, but apart from this, the component values in the equivalent circuit, Fig.1c, are $R_a(\omega) = \hat{R}_a(\sin x/x)$ where $\hat{R}_a \sim$ $k_{eff}^2/\omega_o C_s$ and $x = N\pi(\omega-\omega_o)/\omega_o$.   (The radiation impedance also contains a reactive component which is neglected since it vanishes at ω_o).   The electrical bandwidth is approximately $\Delta\omega/\omega_o \sim (\omega_o C_o \hat{R}_a) \sim Nk_{eff}^2$.   Equating this to the acoustic bandwidth, 1/N, gives for the optimum instantaneous bandwidth, $\Delta\omega/\omega_o \sim 1/N \sim k_{eff}$.   Values of this optimum bandwidth and other important parameters of a piezoelectric surface wave medium are listed in Table II.

The most important parameters of a surface wave material for transduction and propagation are:

(i)   The value of k_{eff}^2.   This is difficult to calculate for surface waves but is typically an order of magnitude smaller than for bulk waves in the same material. Nevertheless the instantaneous bandwidths are comparable, while the surface wave structure is far easier to fabricate.   For ordinary delay lines a high value of k_{eff}^2 is desirable but for certain tapped delay line applications (discussed later) too high a value can be embarrassing.

(ii)   The acoustic losses in the piezoelectric are important for high frequency work.   (In bulk wave transducers

this is much less important since the ultrasonic wave
only travels through the thickness t ∿ $\lambda/2$ of the
transducer material).

(iii) The surface wave velocity, c, should be high for high-
frequency work since (at a given frequency) the
wavelength is proportional to c.   In contrast, for
long delay times a small value of c is desirable.

(iv) The temperature coefficient of the delay time can be
important in surface wave tapped delay line filters.
At present, quartz is the only crystal known to have
zero temperature coefficient cuts (Schultz et al. 1970;
Lewis et al. 1971).

C.   THE ROLE OF OXIDES AS TRANSDUCER MATERIALS

It is clear from Tables I and II that the only commonly
used piezoelectric transducer material which is not an oxide
is CdS, and even this is rapidly being replaced by ZnO.
Thus the piezoelectric transducers are dominated by the
oxides.   The most potent for both bulk wave and surface
wave use is $LiNbO_3$ (Warner and Meitzler, 1968; Smith et al.
1969).   However, $LiNbO_3$ is currently used very little
because it is expensive, not too readily available, and
somewhat brittle.

For bulk wave transduction at low frequencies quartz
and poled ceramics are often used, but these are not satisfactory for higher frequency work primarily through bonding
difficulties.   Above ∿ 1 GHz, sputtered ZnO transducers
dominate.

For surface wave work at frequencies below ∿ 50 MHz a
poled ceramic (PZT) can be used but it is somewhat lossy.
At frequencies of ∿ 100 MHz (Y-cut, x-propagating) quartz is

commonly used while delay lines at 1 GHz have employed
LiNbO$_3$.   While quartz is commonly used because of its reproducibility, cheapness, etc. the value of k^2_{eff} is not
quite as high as desired, and a replacement could be most
useful.   An attractive possibility is the use of sputtered
ZnO layers for the excitation of surface waves on non-
piezoelectric media.   Waves at $\sim$ 100 MHz have been excited
using the I.D.T. on 3μ layers of ZnO deposited on the
following substrates: glass, fused silica, silicon, spinel,
sapphire, YIG and Gd-Ga-garnet.   The use of a silicon substrate seems particularly attractive because of the
possibility of integrating surface wave components with
microelectronics in an entirely planar structure.   As
discussed later a variety of signal processing functions can
be achieved with surface wave components.

D.   BULK WAVE DELAY MEDIA

The most desirable physical properties of the delay
medium are usually: (a) Compatibility of the acoustic
impedance, Z, with that of the transducer medium. (b) Suitable hardness for machining and polishing to optical
tolerances.   (At 9 GHz the acoustic wavelength is of the
same order as the wavelength of light, so that the delay
line tolerances are similar to laser tolerances). (c) Low
ultrasonic propagation losses.

For operation at microwave frequencies at room temperature (c) is the over-riding consideration, but fortunately
many oxides come close to satisfying all these requirements.
For quartz, typical losses are 4 dB/cm at 1 GHz (1cm of
travel corresponds to $\sim$ 2μs delay) rising to $\sim$ 360 dB/cm at
9 GHz. Obviously the latter loss cannot be tolerated in

devices, and for this reason much work has been performed to
understand the intrinsic attenuation mechanism in dielectrics
and to find materials with lower losses than quartz.   It is
therefore appropriate to discuss this attenuation mechanism.
Good single crystals of dielectrics have the lowest losses
as they do not suffer attenuation due to dislocations or
free electrons (in metals) or scattering from grain boundaries (in polycrystalline materials).   The only other
materials which appear promising are glasses.   Unfortunately
fused silica, which has the lowest losses of any glass, has
its attenuation dominated by a relaxation mechanism which is
a result of the disorder in a glass.

It is found that the attenuation results from the
interaction of the ultrasonic wave with thermal phonons.
This interaction occurs through the lattice anharmonicity,
i.e. because the potential energy density, ϕ, is not quite
quadratic in the strains, S, but also contains cubic and
higher order terms,

$$\phi \;=\; \frac{1}{2} c_{ij}\, S_i\, S_j + \frac{1}{6} c_{ijk}\, S_i\, S_j\, S_k \qquad \ldots(1)$$

where c_{ij} and c_{ijk} are respectively the second and third
order elastic constants.   The lattice anharmonicity is also
responsible for thermal expansion (which can be regarded as
an interaction between the thermal phonons and a static
strain) and thermal resistance (which arises from psuedo-
momentum-destroying phonon interactions, e.g. Umklapp
processes).   Thus it is not surprising to find a close
relationship between these quantities and the ultrasonic
attenuation.

An equivalent view of the lattice anharmonicity is
that the stress-strain relationship is non-linear, which

leads to mixing of the ultrasonic and certain thermal phonons generating new thermal phonons with sum and difference frequencies (and wave vectors):

$$\omega' = \omega(\text{thermal phonon}) \pm \omega(\text{ultrasonic})$$
$$\underline{k}' = k(\text{thermal phonon}) \pm k(\text{ultrasonic})$$

$$\ldots(2)$$

but only if $(\omega',\underline{k}')$ is itself an allowed phonon.

This picture is correct at low temperatures where the thermal phonons have well-defined ω and $\underline{k}$. At higher temperatures (including room temperature) the thermal phonons have such short lifetimes, τ, (because of interactions amongst themselves) that the condition $\omega\tau < 1$ is satisfied where ω is the ultrasonic frequency. An approximate value of τ can be obtained from the expression for the thermal conductivity,

$$K = \frac{1}{3} C \bar{c}^2 \tau \qquad \ldots(3)$$

where C is the specific heat/unit volume and $\bar{c}$ an average thermal phonon velocity. In the $\omega\tau < 1$ regime the uncertainty in the thermal phonon frequencies $\Delta\omega \sim 1/\tau$ exceeds the ultrasonic frequency, so that the conservation conditions, Eq.(2), are no longer relevant, and the ultrasonic phonon can interact with any thermal phonon. In these circumstances the attenuation is caused by a relaxation mechanism which arises as follows. The ultrasonic wave modifies the frequencies of the thermal phonon wave packets. The delayed relaxation of this modified phonon distribution back towards equilibrium with a time constant, τ_1, causes an ultrasonic loss of the form

$$Q^{-1} = \left(\frac{\Delta c_{ij}}{c_{ij}}\right) \frac{\omega\tau_1}{1 + \omega^2\tau_1^2} \text{ or } \alpha = \frac{\Delta c_{ij}\omega^2\tau_1}{2\rho c^3(1 + \omega^2\tau_1^2)} \qquad \ldots(4)$$

where ρ is the density, $c_{ij} = \rho c^2$ is the elastic constant and Δc_{ij} is the small difference between the high and low frequency elastic constants ($\omega \gg 1/\tau_1$ and $\omega \ll 1/\tau_1$).

The modified thermal phonon frequency, ω_i^i, is given by

$$\omega_i = \omega_{i_o} \ (1 - \gamma_j^i \ S_j) \qquad \qquad ...(5)$$

where γ_j^i is a Gruneisen number, closely related to the Gruneisen constant, γ. It is interesting to note that values of the γ_j^i can be measured from the change in frequency of an ultrasonic wave under a static stress. For several crystals (including MgO, LiF, KCl, NaCl, YIG, Si, Ge, Al_2O_3 and SiO_2) sufficient measurements have been made to evaluate all the third order elastic constants and so to calculate any required γ_j^i. This is useful for detailed calculations of the ultrasonic attenuation (and also the expansion coefficient), but for many purposes it is sufficient to note that the γ_j^i are of order unity.

An easily visualised example of this type of mechanism is the thermoelastic attenuation mechanism. A longitudinal ultrasonic wave causes alternate regions of the lattice separated by $\lambda/2$ to be compressed and rarefied. Normally the compressed regions are heated (equivalently, there is a net rise in the frequencies of the thermal phonons in the compressed region) while the rarefied regions are cooled. The irreversible heat flow between these regions by thermal conduction causes the ultrasonic attenuation. For the thermoelastic mechanism Δc_{ij} is the difference between the adiabatic and isothermal elastic constants and, for an isotropic material is given by

$$\Delta c_{11} = c_{11}^{ad} - c_{11}^{is} = \gamma^2 CT \qquad \qquad ...(6)$$

where T is the absolute temperature and the volume Gruneisen constant $\gamma = \langle \gamma_j^i \rangle = 3\alpha'B/C$ (α' = linear expansion coefficient, B = bulk modulus).   In practice this mechanism accounts for only a few percent of the attenuation of longitudinal ultrasonic waves while it is inoperative for shear waves.   The reason for this can be seen from Fig.3 which shows a shear wave propagating on a two-fold axis.   The regions I and II separated by $\lambda/2$ suffer opposite shears but are in identical states.   Therefore there cannot be any temperature difference between these regions, and there cannot be any thermoelastic losses.   However, Fig.3b shows how another loss mechanism can arise.   The shear suffered by region I can be represented by an expansion and a contraction in two directions at right angles, as shown. Consequently the two thermal phonons α and β suffer equal and opposite frequency changes.   Thus while there is no net frequency change in region I (and no temperature change) there is a non-equilibrium distribution of thermal phonons within this region.   This non-equilibrium distribution returns to equilibrium through phonon-phonon interactions with a time constant of the order of τ in Eq.(3).   This mechanism, Akhiezer (1939), appears to account for the measured attenuation in many materials for which the second and third-order elastic constants are known (Mason and Bateman, 1966; Lewis, 1968).

For this mechanism $\Delta c_{ij} = \delta^2 CT$ where

$$\delta^2 = \langle (\gamma_j^i - \langle \gamma_j^i \rangle)^2 \rangle \qquad \ldots (7)$$

is the mean square deviation of the γ_j^i from $\langle \gamma_j^i \rangle$, and is again of order unity.   However, for certain shear waves (particularly those propagating on the <100> axis of cubic

TABLE III

Material	Wave l=longitudinal t=transverse f.t=fast transverse	Acoustic velocity c (m/s)	Acoustic impedance Z (10^6 kg/m^{2}s)	Measured attenuation (dB/cm at 9 GHz)	Calculated attenuation of longitudinal waves from Eq(8) with $\delta^2=1$ (dB/cm at 9GHz)	Debye Temperature (°K)
Al_2O_3	l(c-axis)	11000	44	15	19	1040
	t(c-axis)	6300	25			
$MgAl_2O_4$	l<100>	8800	32	36	34	950
	t<100>	6500	23	10		
$Y_3Al_5O_{12}$	l<100>	8600	40	26	14	750
	t<100>	5000	23	25 – 30		
SiO_2	l(x-axis)	5700	15	360*	220	586
	f.t(x-axis)	5100	14	100*		
Fused silica	l	5900	13	2000*		
	t	3700	8	1600*		

* Extrapolated from lower frequency measurements.

crystals) many of the γ_j^i vanish by symmetry and δ^2 is reduced
by up to an order of magnitude. From this discussion the
Akhiezer attenuation coefficient is (for $\omega\tau \ll 1$)

$$\alpha(cm^{-1}) = \frac{\delta^2 C T \omega^2 \tau}{2\rho c^3} \simeq \frac{3\delta^2 K T \omega^2}{2\rho c^3 \bar{c}^2} \qquad ...(8)$$

Oliver and Slack (1966) have used this expression with $c = \bar{c}$
to show how the average attenuation coefficient varies with
other crystal properties. Although many of these properties
are inter-related, in general the requirements for low
losses are (a) a high Debye temperature, θ_D (light atoms)
and (b) complex structures. The former requirement
essentially comes from the velocity factor in the denominator
of Eq.(8) (since materials with high θ_D have high velocities).
The second requirement comes from the fact that complex
crystals have low thermal conductivities. Fortunately
several crystals satisfy both these requirements, mainly the
hard oxides, Table III.

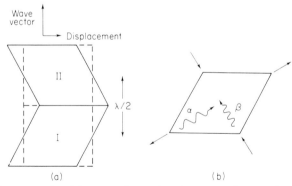

*Fig.3. a) Illustrating the deformation under a
shear wave, b) shows how two thermal phonons,
α and β, undergo equal and opposite frequency
changes under the shear.*

E.   ROLE OF OXIDES AS DELAY MEDIA

Table III shows that the oxides dominate the low-loss media.   There are a few other materials which promise to be low-loss, e.g. diamond and β-boron, but neither of these is available in suitable form even for proper attenuation measurements.   Of the low-loss media, YAG, spinel and Al_2O_3 are all available as large Czochralski grown crystals and have been used in delay lines.   Spinel is potentially the best medium but the low-loss shear wave is not compatible with sputtered ZnO transducers which only generate longitudinal waves efficiently at 9 GHz.   It is, however, possible to mode-convert longitudinal waves to shear waves on reflection from an appropriate surface (Lean and Shaw, 1966) and delay lines employing mode-conversion have been produced using YAG and spinel.   The complexity of the mode-converter has meant that the simpler combination of ZnO transducers on Al_2O_3 has largely been employed (Olson, 1970) but the performance is inferior to that obtainable with spinel.

III.   MAGNETOELASTIC DELAY LINES

In a ferrimagnetic insulator each unit cell possesses a net spin magnetic moment, and these moments are coupled by short-range exchange forces and long-range dipolar forces such that the state of lowest energy has the moments aligned. Deviations from the aligned ground state can be resolved into spin waves or magnons just as deviations in the positions of atoms from a regular lattice are resolved into sound waves or phonons.   In principle these spin waves can be employed in delay lines analogously to sound waves, but in practice only one material, YIG ($Y_3Fe_5O_{12}$) has been found suitable.

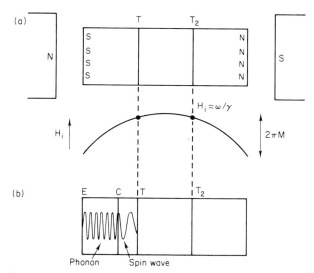

Fig.4.   Illustrating the internal magnetic field variation and magnetoelastic wave-function in an axially magnetised YIG rod.

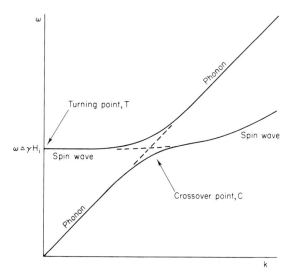

Fig.5.   Dispersion curves for spin waves and phonons in YIG.

The reason is simply that in all other materials the spin
waves are too heavily damped to propagate for useful periods.
This can be seen from the approximate expression for the
lifetime, τ, of a magnetic excitation, $\tau^{-1} = \Delta\omega = \gamma\Delta H$ where
ΔH is the ferrimagnetic resonance linewidth and γ the gyromagnetic ratio.   For YIG, $\Delta H \sim 0.1$ Oe, giving $\tau \sim 0.5$ μs,
while for most other ferri/ferromagnets, $\Delta H \sim 100$ Oe, giving
$\tau \sim 10^{-9}$μs.   The low magnetic losses in YIG are not directly
related to its being an oxide, although, being a good insulator eliminates eddy-current damping of the spin waves, and
the oxygen ions may take part in the exchange coupling
through the super-exchange mechanism (Anderson, 1963).
Probably the most important single reason for the low magnetic losses in YIG is that all the Fe ions are in S-states.

One of the most interesting spin wave effects in YIG
is the propagation of magnetoelastic waves in an axially
magnetised YIG rod (Schlomann et al. 1965).   The rod is
typically 1cm long and 3mm in diameter.   In such a rod the
demagnetising field is non-uniform and varies from $\sim$ zero at
the centre to $\sim 2\pi M$ at the ends.   The resulting internal
field distribution, H_i, is shown in Fig.4a.   The following
is necessarily a simplified account of the propagation.
The spin wave dispersion relationship is of the form

$$\omega = \gamma H_i + \gamma D k^2 \qquad \qquad ...(9)$$

where D is an exchange energy constant.   By adjusting H_i
this can be made to cross the phonon dispersion curve,
$\omega = ck$.   In the presence of magnetoelastic coupling
(magnetostriction) these dispersion curves split as shown in
Fig.5.   It is then possible to excite a spin wave at T
which traverses the upper branch of the dispersion curve and

converts to an elastic wave at the crossover point, C. Experimentally this is achieved by applying a microwave pulse to a loop situated near one end of the rod.   The spin wave travels down the magnetic field gradient from T to C, and thence to the end-face E as an elastic wave.   On reflection from E it undergoes the reverse sequence of events and radiates a microwave echo on its return to T.   Although it is convenient to picture this as a traversal of the dispersion curve, in reality the frequency, ω, remains fixed and the dispersion curves move because of the variation of H_i with position.   The round-trip delay time varies with the applied magnetic field since the latter determines the position of T and hence the pathlength from T to E.   This delay typically varies from 2 to 6 μs for a change of 100 Oe, which can be useful in devices.   The propagation is also dispersive since the position of T is determined by $H_i = \omega/\gamma$. Typically the delay varies by a few μs over 100 MHz, which is a suitable variation for use in pulse compression radar.

In practice, spin wave losses are approximately $(2f + 3)$dB/μs (Strauss, 1965) where f is the frequency in GHz, and this limits the use at room temperature to $f \stackrel{\sim}{<} 3$ GHz and delays of a few μs.   For completeness, it should be noted that recent work has shown that the spin wave is not excited directly at T, but first travels from the end-face to T as a magnetostatic (long wavelength dipole-coupled) spin wave (Addison et al. 1968).

IV.   SOME APPLICATIONS OF HIGH FREQUENCY DELAY LINES

Delay lines are used for a variety of functions, some of which are noted below.   Bulk wave delay lines are somewhat restricted in application because their sole purpose is

M. F. Lewis

delay, i.e. they have only one input and output port.   In
contrast, surface wave delay lines can be monitored at any
point in their path, and this allows them to be used for
many signal processing functions, in particular as tapped
delay line filters.

The simplest use of delay lines is in radar systems
where information is stored from one echo to another for
comparison or correlation purposes.   One particular application is in range calibration, where a microwave ultrasonic
delay line gives a delayed echo corresponding unambiguously
to a predetermined range.   Where longer delays are required
(e.g. $\sim$ 1 ms for pulse-to-pulse comparison in radar) this
can be achieved with bulk waves using a multiple reflection
technique or with surface waves using a helical path on the
surface of a rod.

Another application of microwave ultrasonic delay
lines is in a recirculating store with rapid access.   For
example, a 1 μs delay line with 1 GHz bandwidth can accommodate 1000 bits with access time of $\sim$ 1 μs.   This compares
favourably with a conventional low frequency line of the
same capacity but with a bandwidth of $\sim$ 1 MHz and delay and
access times of $\sim$ 1 ms.

Dispersive delay lines are of use in pulse compression
radar.   In this technique a radar pulse is transmitted with
long duration, τ_1, and linear frequency sweep Δf.   On
returning from the target this frequency modulated pulse is
compressed in a dispersive delay line to a width $\tau_2 \sim 1/\Delta f$,
where $\tau_1/\tau_2 \sim 100$.   The primary use of this technique is to
obtain a higher transmitted pulse energy than if the trans-
mitter ran for period τ_2 (at constant frequency).   It
cannot be allowed to run for period τ_1 at constant frequency

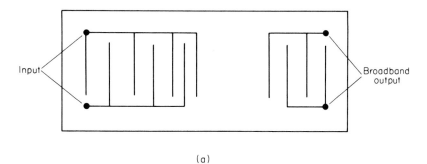

(a)

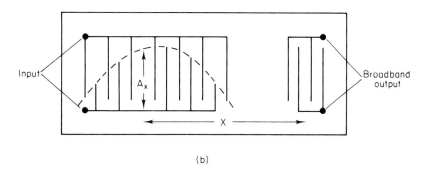

(b)

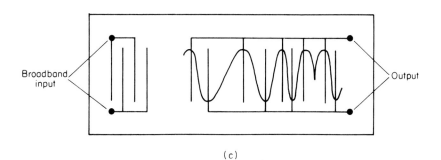

(c)

Fig.6. Schematics of three surface wave devices
a) dispersive delay line, b) filter, c) correlator

without degrading the range resolution.   A second advantage
of this technique is the relative immunity to jamming.   One
suitable dispersive delay line which operates at the radar
(microwave) frequency is the YIG magnetoelastic line mentioned previously.   This has about the required dispersion
(a few μs over 100 MHz) but has not found great application,
largely because of the difficulty of making reproducible
material.

A second dispersive delay line can be made using
surface waves, where one of the I.D.T.'s has a graded period-
icity, Fig.6a.   In this case the surface wave is excited in
that region of the I.D.T. where the wavelength matches the
transducer period, and the delay time is thus dispersive.
With this technique, pulse compression filters have been
made (using quartz) with compression ratios of $\sim$ 100 at a
centre frequency of $\sim$ 60 MHz.

Surface waves can be used to make filters and cor-
relators, and form the first really viable form of the
tapped delay line filter discussed by Kallman (1940).   By
tapping and summing signals at various points a variety of
filter characteristics can be produced.   A simple illustra-
tive example is shown in Fig.6b, which comprises an amplitude
weighted input I.D.T. and broadband output I.D.T.   For an
input of frequency ω, the output is given by

$$O(\omega) = \Sigma \, A(x) \, \exp \, (-jkx) \qquad \qquad ...(10)$$

where $\omega = ck$.   This is the Fourier transform of the amplitude weighting function.   Consequently the design of
filters is fairly trivial, but a number of complications
arise when the filter is being designed to a tight specification.

The final application mentioned here is the surface
wave correlator, used for detecting signals with a known
coded waveform in the presence of noise or jamming. The
corrupted received signal is fed into the broadband input of
Fig.6c and runs through the output transducer, which has
been designed to match the signal waveform as shown. Only
at the instant shown is a large output pulse obtained; at
other times (or for other waveforms) the signals from each
finger pair in the output I.D.T. add incoherently to give a
smaller output.

REFERENCES

Addison, R.C., Auld, B.A. and Collins, J.H. (1968) J. Appl.
    Phys. 39, 1828.

Akhiezer, A. (1939) J. Phys. Moscow 1, 277.

Anderson, P.W. (1963) "Magnetism" Vol.1 (G.T. Rado and
    H. Suhl, eds.) p.25, Academic Press, New York.

Jenkins, F.A. and White, H.E. (1957) "Fundamentals of
    Optics", 3rd ed. McGraw-Hill.

Kallman, H.E. (1940) Proc. I.R.E. 28, 302.

Lean, E. and Shaw, H.J. (1966) App. Phys. Lett. 9, 372.

Lewis, M.F. (1968) J.A.S.A. 44, 713.

Lewis, M.F., Bell, G. and Patterson, E. (1971) J. Appl.
    Phys. 42, 476.

Mason, W.P. and Bateman, T.B. (1966) J.A.S.A. 40, 852.

Oliver, D.W. and Slack, G.A. (1966) J. Appl. Phys. 37,
    1542.

Olson, F.A. (1970) Microwave J. 13, 67.

Schlomann, E., Joseph, R.I. and Kohane, T. (1965) Proc.
    I.E.E.E. 53, 1495.

Schulz, M.B., Matsinger, B.J. and Holland, M.G. (1970)
J. Appl. Phys. <u>41</u>, 2755.

Smith, W.R., Gerard, H.M., Collins, J.H., Reeder, T.M. and
Shaw, H.J. (1969) I.E.E.E. Trans. Mic. Theory and
Techniques <u>MTT-17</u>, 856.

Strauss, W. (1965) Proc. I.E.E.E. <u>53</u>, 1485.

Warner, A.W. and Meitzler, A.H. (1968) Proc. I.E.E.E. <u>56</u>,
1376.

White, R.M. and Voltmer, G.W. (1965) App. Phys. Lett.
<u>7</u>, 314.

OXIDE GLASSES

J.A. SAVAGE

Royal Radar Establishment,
Malvern, Worcestershire

I. INTRODUCTION

II. THEORIES OF GLASS FORMATION

III. GLASSES AND GLASS FORMING SYSTEMS

    A. SILICATE GLASSES

    B. PHASE SEPARATION AND CONTROLLED
       CRYSTALLISATION IN SILICATE SYSTEMS

    C. BORATE, PHOSPHATE, TELLURITE, GERMANATE,
       VANADATE AND ALUMINATE GLASSES

IV. GLASSES TO MEET THE DEMANDS OF MODERN TECHNOLOGY

    A. SEMICONDUCTING GLASSES

    B. OPTICAL GLASSES

I. INTRODUCTION

A glass is defined as an inorganic product of fusion
which has cooled to a rigid condition without crystallising
(Morey, 1945). Glasses differ from crystalline substances
in many of their physical and chemical properties. The
chemical composition of a multi-component glass cannot be
expressed by a stoichiometric formula, because within any
glass forming system the composition can usually be

99

continuously varied, yielding a family of glasses possessing
a range of different physical properties.   This is one of
the most useful characteristics of glasses in that it allows
a composition to be optimised to suit a particular requirement.   However, there is one very important property which
is rather insensitive to changes in composition within a
glass forming system - mechanical strength.   The tensile
strength of glass is about 10-15 times less than the
compressive strength, so that glass usually fails in tension.
Cracks are initiated at the glass/air surface and propagate
rapidly, leading to component failure.   The mechanical
strength of glass is therefore dominated by its surface
condition, and depends to a greater degree on its resistance
to abrasion and chemical attack than to intrinsic strength
variations brought about by changes in chemical composition
within a particular glass forming system (Griffith, 1920).
Homogeneous glasses are isotropic and on cooling from the
molten state their properties change continuously with temperature, thus allowing articles possessing uniform physical
properties to be made by simple moulding techniques.
Glasses do not possess a melting point and, in contrast to
crystalline materials, their structural lattice or network
is irregular and aperiodic.   By definition metastable
glass-melts transform to solid glasses at the glass transition temperature, T_g (Jones, 1956).   The value of T_g
depends on the chemical composition of the glass and upon
the rate at which the metastable melt is cooled or quenched.

The cooling process can be understood by reference to
Fig.1 which shows how the volume of a melt changes when its
temperature is decreased until the solid state is reached.
At high temperature, above the melting point, T_m, the liquid

is stable but once this liquid is super-cooled below its
melting point it becomes metastable. As the temperature is
further decreased the structural configuration of the metastable liquid follows the temperature and is in equilibrium
with it. However, as the temperature of the metastable
liquid approaches T_g, the viscosity becomes increasingly
high so that the structural configuration takes more time to
adjust to its equilibrium value, until at T_g it can no
longer do so rapidly. Hence at normal cooling rates a
certain structural configuration, which would be in equilibrium with some temperature in the region of T_g, is
'frozen' into the now solid metastable glass. It is
important to realise that T_g is not a unique temperature for
any particular glass composition but depends on the previous
thermal history of the glass and to a small extent upon the
method of measurement. On fast cooling a structural configuration representing a higher temperature is 'frozen'
into the glass, T_{gf} in Fig.1. On slow cooling a structural
configuration representing a lower temperature is 'frozen'
into the glass, T_{gs} in Fig.1. T_g values quoted in the
literature usually represent the lower value. Fig.1 may be
divided into three important temperature regions: (1) below
T_g the material is described as a solid metastable glass
having a geological lifetime, (2) between T_g and T_m the
material is described as a metastable glass melt with a pre-
crystallisation lifetime of minutes or hours, (3) above T_m
the material is described as a stable melt whose lifetime is
very long. It is between T_m and T_g that all shaping operations to form glass components are carried out.

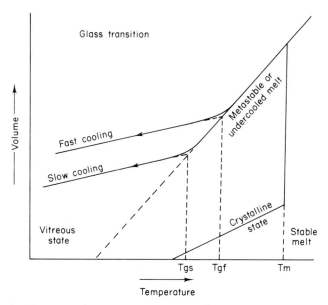

*Fig.1. A volume-temperature plot for glasses
subjected to fast and slow cooling rates.*

Having formed a component and cooled it to a temperature just above T_g it is necessary to hold it at this
temperature for annealing.   Rapid cooling of a component
results in large internal stresses causing spontaneous
fracture at room temperature.   During annealing a component
is brought to thermal equilibrium and hence to a uniform
structural configuration which is frozen into the glass at
T_g.   Thereafter, it is cooled slowly to avoid large temperature gradients.   If large gradients are used the required
differential contraction could not occur because viscous
flow or structural configuration adjustment is no longer
possible at temperatures well below T_g and 'temporary stress'
would appear in the component.   Temporary stress in a solid
glass results from a temperature gradient and is removed

when the temperature gradient is removed.   If a component
is made and held just above T_g in a temperature gradient so
that the structural configuration can adjust to the gradient
and then be 'frozen' into the glass at T_g, no stress exists
in the solid glass on cooling as long as the temperature
gradient is maintained.   On cooling to room temperature
permanent stress is introduced because the required structural adjustments cannot occur at low temperature on removal
of the temperature gradient.   Permanent stress in a glass
therefore arises from temperature gradients within the glass
when cooled through the glass transition temperature.   This
stress can only be removed by isothermally re-annealing the
glass above T_g.

The viscosity ranges associated with the cooling
process are often quoted and these are defined in Table I.

TABLE I

POINT OR RANGE IN THE COOLING PROCESS	VISCOSITY (POISES)
Working range for shaping components	$10^4 - 10^6$
Littleton softening point (Fibre extension method)	$10^{7 \cdot 6}$
Dilatometric softening point, M_g	10^{11}
Annealing point	10^{13}
Glass transition temperature, T_g	$10^{13 \cdot 3}$
Strain point	$10^{14 \cdot 5}$

II.   THEORIES OF GLASS FORMATION

Single component glasses can be made from SiO_2, GeO_2,
B_2O_3, P_2O_5 and As_2O_3.   Each of these oxides can be melted

with one or more non-glass forming oxides and cooled to form
a glass with modified properties.   A second group of oxides,
TeO_2, SeO_2, MoO_3, WO_3, Bi_2O_3, Al_2O_3, Ga_2O_3 and V_2O_5 do not
form single component glasses but will do so when melted
with another suitable non-glass forming oxide.   The second
group may be described as intermediate oxides.

Zachariasen (1932) postulated a set of rules for glass
formation based on structural energy considerations as
follows: (1) no oxygen atom may be linked to more than two
atoms A,   (2) the number of oxygens surrounding A must be 3
or 4,   (3) the oxygen polyhedra share corners with each
other not edges or faces,   (4) at least three corners in
each polyhedron must be shared.

Warren (1937) determined the structure of simple
silicate glasses (SiO_2 and Na_2O-SiO_2) by X-ray diffraction
techniques and proposed a model in agreement with Zachariasen.
The structural model was based on a continuous random network
of SiO_4 tetrahedra with the sodium atoms fitting into large
interstices within the network.   In fused silica all the
oxygens are bonded to two silicon atoms, but in sodium
silicate glasses extra oxygen atoms are introduced by the
Na_2O.   Thus, some oxygens in sodium silicate are bonded to
two silicon atoms as before and are called 'bridging oxygens',
while others are bonded only to one silicon atom and are
called 'non-bridging oxygens'.   The charge neutrality is
maintained by a sodium atom in an interstice in the network.
This Zachariasen-Warren random network theory has had wide
acceptance and it is only recently, with the discovery of
new glasses which do not obey the Zachariasen rules, that
the limitations of the theory have been recognised.   However,
it is still valid for the simple alkali silicate compositions.

Stanworth (1946) established a correlation between the type of bonding in an oxide material and its ability to form a glass. He attempted to quantify the correlation by using Pauling's values for electronegativities of the elements, and found that the electronegativity difference between oxygen and each glass forming element ranged from 1.8 to 2.1. He also found a value of 2.1 for tellurium and oxygen, thus successfully predicting glass formation in TeO_2-XO systems. This criterion cannot be applied universally because SnO_2 and Sb_2O_3 do not form single component glasses although they have the required electronegativity difference values.

Sun (1947) suggested a correlation between bond-strength of the X-O bond in an oxide XO and glass formation. All glass forming oxides have high single bond-strengths of $\sim$ 90-100K cal/mole. However, V_2O_5 and Sb_2O_3 fall within this range but do not form single component glasses. Rawson (1956) suggested that the thermal energy available to break the bonds should also be considered and arrived at a new controlling parameter, B_{X-O}/T_m, by dividing the bond-strengths by the melting temperature. Again V_2O_5 proves an exception.

Turnbull and Cohen (1958) proposed a kinetic theory of glass formation in which limiting rates for the processes of nucleation and crystal growth in liquids are established for glass formation to occur. This theory is difficult to apply to new systems in order to predict glass formation because parameters such as the chemical potential and surface energy of liquid crystal interfaces required in the calculations are unknown.

There are, as yet, no universal criteria for glass formation, but the lower the melting point or liquidus

temperature of a material in relation to its bonding energy
the better the chance of glass formation. Other effective
factors in promoting melt supercooling and glass formation
are the existence of a primary crystalline phase of a
different composition to the average melt composition, so
that a good deal of atomic motion and diffusion is necessary
before nucleation and crystal growth can occur, and a high
viscosity at the liquidus to make these diffusion processes
more difficult. Therefore, in attempting to predict glass
formation in a new system, a knowledge of the phase diagram
and melt viscosities is a minimum requirement.

III. GLASSES AND GLASS FORMING SYSTEMS

In order of importance, the oxide glass forming systems
are (1) silicate, (2) borate, (3) germanate, phosphate,
tellurite, aluminate, gallate and vanadate.

A.   SILICATE GLASSES

Silica is the only single component glass which is made
on a commercial scale (Bruckner, 1970, 1971). Its usefulness stems from a high softening temperature ($T_g \sim 1100^\circ C$),
excellent visible and UV transparency and good chemical
durability. It is more difficult to manufacture than other
glasses because it melts at temperatures in excess of $1723^\circ C$.
Silica was first made by flame fusion and later by electric
fusion in carbon or molybdenum crucibles. More recently it
has been manufactured synthetically by the hydrolysis of
$SiCl_4$ in an oxy-hydrogen flame, or by the pyrolysis of $SiCl_4$
in a water free plasma flame. The properties and impurity
content (metallic and hydoxyl) depend on the method of
manufacture. Electrically fused silica may contain of the

order of 100 ppm of metallic impurities while the synthetic
silica produced by hydrolysis may contain of the order of
1000 ppm of OH.   Care must therefore be taken in choosing a
particular grade of SiO_2 to suit a technical requirement.

Other high silica glasses have also been produced by
the vapour process.   A TiO_2 modified silica glass has been
developed with zero thermal expansion between $0^{\circ}C$ and $200^{\circ}C$
for use in the construction of telescope mirrors.   The use
of the vapour technique for multi-oxide glasses is limited
because of the differential vapour pressure of the
components.   Furthermore, multi-oxide glasses possess lower
softening temperatures and are therefore more easily made by
a melting process.

Many multi-oxide silicate glass compositions exist,
but the three glasses made in the greatest quantities are
soda-lime-silica, pyrex, and lead silicate.   A typical
soda-lime-silica composition is, $SiO_2(72.1\%)$ $Al_2O_3(1.8\%)$ MgO
and $CaO(9.8\%)$ $BaO(0.3\%)$ Na_2O and $K_2O(15.6\%)$ B_2O_3, SO_3 and
$F_2(0.4\%)$ by wt.   This type of material is made in large
quantities in tank furnaces and processed by automatic
machinery to form containers, tubing and sheet glass.
Pyrex was developed to provide a glass of better chemical
durability and thermal endurance than soda-lime-silica glass.
The average chemical composition of pyrex is $SiO_2(80.9\%)$
$B_2O_3(12.9\%)$ $Al_2O_3(1.8\%)$ $Na_2O(4.4\%)$ by wt. and it is from
this type of glass that most heat resisting articles are
manufactured for domestic and scientific purposes.   The
need for a glass of higher refractive index than soda-lime-
silica for optical purposes resulted in the development of
lead silicate glasses of which a typical composition is
$SiO_2(65\%)$ $PbO(20\%)$ $Na_2O(7.5\%)$ $K_2O(7.5\%)$ by wt.   This

material is now used mainly to produce decorative lead
crystal glass for the domestic market. From these three
basic silicate glass compositions a whole range of glasses
for technical purposes has been developed, including low
dielectric loss glasses for use in the electrical industry,
sealing glasses of specific expansion coefficient used in
the electronics industry, special glasses for thermometer
tubing and glass electrodes, and 'neutral' glasses for
pharmaceutical use.

   Glasses manufactured for optical purposes are mostly
silicates ranging in refractive index from 1.45 to 1.8 and
v_d value 70 to 25, where v_d is n_d-1/n_f-n_c, the reciprocal
of the dispersion in relation to the refraction. A low v
value is characteristic of a glass with a high dispersion
relative to its refractive index, and this type of glass is
called a flint. A low dispersion relative to the refractive index is characteristic of a high v value glass, and
this is called a crown ($v_d \geqslant 50$). Extra care has to be
taken in homogenising and annealing optical glasses in order
to obtain uniform properties. With some compositions it is
possible to reduce refractive index variations to $\sim \pm 1 \times 10^{-6}$.

B.   PHASE SEPARATION AND CONTROLLED CRYSTALLISATION
      IN SILICATE SYSTEMS

   The useful region of glass formation in several binary
and ternary borate and silicate systems is limited by the
separation of the melt into two liquids of different composition (Rawson, 1967). Glasses made from separated melts
have either a layer structure or are opalescent in appearance. When the critical temperature at which phase
separation takes place is below the liquidus temperature,

the glass produced on supercooling is phase-separated on a
very fine scale, detectable only by electron microscopy
techniques.   This effect is used in the manufacture of
Vycor, a 94% SiO_2 glass marketed in the USA.   Articles are
made by the normal glass working techniques and then heat
treated at a temperature in the immiscibility region but not
high enough to cause deformation.   The minute droplet glass-
phase within the parent glass grows until there is a
substantial interconnected droplet phase present within a
'honeycomb' of the main glass phase.   This process occurs
within a certain composition range in the sodium-borosilicate
system.   The secondary droplet phase is sodium borate and
the primary phase is almost pure SiO_2.   A typical initial
glass composition might be SiO_2(75%) B_2O_3(20%) Na_2O(5%) by
wt. and after components have been heat treated and annealed
they are acid leached to remove the sodium borate phase.
When dry the primary 'honeycomb' phase has a composition
SiO_2(94%) B_2O_3(5-6%) Na_2O(0.25%) by wt., and contains pores
ranging from 20 to 40 $\overset{o}{A}$ in size.   The leached components
are then fired between $900^{o}C$ and $1200^{o}C$ to form non-porous
transparent 94% SiO_2 glass components reduced in volume by
about 30%.

Photo processes can occur in silicate glass when it is
doped with certain impurities.   These processes have been
investigated and optimised, and have led to the production
of photo-sensitive glass, photochromic glass and glass
ceramics (McMillan, 1964; Weyl, 1959).   The window panes
of old buildings are often tinted a purple colour.   This is
due to solarisation from the action of sunlight on impurities
in the glass.   In glass Mn^{2+} is colourless but the absorption of UV light causes Mn^{2+} to lose another electron to

become Mn^{3+}, and this ion absorbs visible light thus
resulting in the purple colour of the glass. The electron
is probably trapped by iron impurity according to the equa-
tion $Fe^{3+} + Mn^{2+} \rightarrow Fe^{2+} + Mn^{3+}$. Practical use can be made
of this type of process in the manufacture of photo-
sensitive glass. Cerium oxide and copper oxide impurity
are added to a suitable host glass which is formed into
plates by normal glass working processes. When the glass
is irradiated by UV light, through a mask or photographic
negative, the reaction $Ce^{3+} + Cu^{+} \rightarrow Ce^{4+} + Cu$ occurs in the
unmasked areas. The copper particles so produced act as
nucleating centres for the controlled crystallisation of the
glass during heat treatment, and the detail of the mask or
negative is reproduced in the glass. It has also been
found that the crystallised glass or glass ceramic part of
the composite can be etched in acid at 15 times the rate of
the original glass thus opening the way for chemical
machining. The etched component can be left in the vitreous
state or re-exposed to UV radiation and heat treated to form
a glass ceramic component. Sieves up to 600 mesh screen
size and glass ceramic fluid logic circuits have been made
using these techniques.

Photochromic glass is a borosilicate glass containing
silver halide crystals 50 - 100 $\overset{o}{A}$ in size separated by $\sim$
500 - 1000 $\overset{o}{A}$ (Megla, 1966; Aranjo, 1968; Garfinkel, 1968).
This crystallite size is considerably smaller than the 1000
$\overset{o}{A}$ diameter size of silver halide crystals in high resolution
photographic emulsions. In these latter materials the
halogen can diffuse from its original site, and is therefore
no longer available for recombination when the irradiation
ceases. However, the glass matrix prevents such diffusion

and the photochromic process is reversible. Activation or
darkening occurs on irradiating with UV or short wavelength
visible light, and fading is enhanced by exposure to heat.
The response and fade time are of the order of 1-2 mins. and
5 mins. respectively, depending upon the intensity of the
irradiation. Because of the slow time constants the main
applications of photochromic materials are opthalmic and
architectural. The original materials were made by adding
silver halide in the glass making process, but in a more
recent development the halide is introduced into the surface
of glass plates by an ion exchange process thus allowing a
greater degree of control over production.

The realisation that photo-sensitive glass could be
converted to a ceramic with a small grain size led to the
development of the relatively new glass ceramic materials
(McMillan, 1964). When an ordinary glass is heat treated
at an appropriate temperature it will crystallise with a
large grain size from a few surface nucleation sites. The
resulting ceramic material is very weak and of little
practical use. Alternatively, if a large number of nucleation sites are provided within the bulk of the material the
crystals which grow on heat treatment are of μm dimensions,
resulting in a fine grained strong ceramic. In photo-
sensitive glass the required nuclei are produced by a photo
process but nucleating agents such as TiO_2 and P_2O_5 can also
be incorporated into the glass and spontaneously precipitated out to form nuclei by suitable heat treatment.
Articles are formed initially by normal glass working techniques and are then subjected to a nucleation heat treatment
of up to 2 hours at a temperature corresponding to a glass
viscosity in the range 10^{11}-10^{12} poise. The articles are

then raised in temperature at ∿ 5°C/min in order to promote
crystal growth on the nuclei produced by the first heat
treatment.   The upper temperature to which they are subjected is usually 50°C below the liquidus of the major
crystalline phase present in the glass ceramic material.
In most cases the ceramic material obtained by this type of
process is white and opaque and the volume change is ∿ 3%.
Thermal expansion may be increased or decreased according to
the initial glass composition and the particular crystalline
phases obtained during heat treatment.   The softening
temperatures are increased by as much as 300-400°C, and the
mechanical strength by 3 to 4 times due to the difficulty of
crack propagation across grain boundaries and the lower
susceptibility of the strength to surface abrasions.   Abrasion reduces the strength of the glass ceramics by no more
than 20%, as opposed to 50% for glasses.   The grain size is
approximately 1 µm and the porosity is essentially zero.
For some applications the properties of glass ceramics are
superior to glasses and to conventional ceramics.   A list
of some glass ceramic systems together with their particular
property of interest is shown in Table II.   Applications
for these materials include cooking ware, sealing and bonding
media, insulators, vacuum tube envelopes, bulk and thick
film capacitors and radomes.

C.   BORATE, PHOSPHATE, TELLURITE, GERMANATE,
      VANADATE AND ALUMINATE GLASSES
      The main use of B_2O_3 is in boro-silicate glasses but
some borate glasses find use in special applications such as
high refractive index rare-earth-borate optical glasses,
sodium resistant alumino-borate glasses for coating the

inside of sodium discharge tubes, and lead borate solder
glasses. A novel use of pure B_2O_3 glass is the Czochralski
growth by liquid encapsulation (Mullin et al. 1964) of some
compound semiconductors such as GaP and InP which have a
high vapour pressure at their melting points.

TABLE II

GLASS CERAMIC SYSTEM	IMPORTANT PROPERTY	MAJOR PHASE
$Li_2O-Al_2O_3-SiO_2$	Low expansion	Beta-spodumene
$MgO-Al_2O_3-SiO_2-TiO_2$	Low dielectric loss, high resistivity	Cordierite, Magnesium titanate, Rutile
$Li_2O-MgO-SiO_2$	High expansion	Quartz, Lithium disilicate
$Li_2O-ZnO-SiO_2$	High strength	Lithium disilicate, Crystobalite, Quartz

Phosphate glasses are not of great technical importance
except that P_2O_5 is used in optical glass manufacture.
Recently, monostable and bistable solid-state switches have
been made from a semiconducting phosphate glass containing a
large percentage of copper oxide in the cupric and cuprous
forms (Drake et al. 1969). These switches are made by
sandwiching a layer of the glass between two copper electrodes and, as fabricated, the resistance of the device is
high. When a certain DC threshold voltage is exceeded the
monostable device switches into a low resistivity state and
remains in this state as long as a certain holding current
is applied to the device. The bistable device remains in
its low resistivity state after switching. The mechanism
of operation of these devices is not as yet understood and
they are still in an early stage of development. If their

operating characteristics, reliability and lifetime are found to be satisfactory it is possible that they might find application in controlling electroluminescent display panels or as memory elements.

Germanate, tellurite, vanadate and aluminate glasses have little commercial importance and have only been manufactured because they exhibit a longer wavelength infrared transmission than silicate or borate glasses. In general, bulk glasses cut off in the infrared in the region of the first overtone bands of their fundamental frequencies. Absorptions arise from the various modes of vibration of the structural groups such as BO_3, BO_4, SiO_4, AlO_4 and TeO_6. In glasses the co-ordination numbers and symmetry of the structural groups are sufficiently similar for the frequencies of the fundamental modes to depend to a great extent on the masses of the atoms present. Hence germanate, tellurite and aluminate glasses transmit to a longer wavelength than silicate or borate glasses. The transmission range of the germanate, tellurite and aluminate glasses is just adequate to cover the 1-6 μm atmospheric window and several glass compositions are manufactured specifically for infrared use. The manufacture of this type of glass requires a specialised technique of vacuum melting or melting and bubbling in a dry gaseous atmosphere in order to remove 'water', which is chemically bonded into the structure of the glass as OH. The OH impurity arises from water in the raw materials, and water vapour in the melting furnace atmosphere. If OH impurity is not removed, an absorption band occurs in silicates at 2.73 μm and renders the material useless for infrared applications. It was the discovery that certain glass compositions could be vacuum melted in carbon crucibles

which allowed oxide glasses to be exploited as infrared
window materials.

IV.   GLASSES TO MEET THE DEMANDS OF MODERN TECHNOLOGY

A.   SEMICONDUCTING GLASSES

Most of the glass compositions already mentioned contain alkali oxides such as Na_2O and K_2O and are therefore
ionic conductors of electricity.   However, glasses free
from these oxides and containing large percentages of multi
valent transition element oxides, such as V_2O_5, Fe_2O_3 or CuO,
are semiconductors of low mobility.   The solid state switch
already discussed is one new application of these semiconductors, but another current interest is their use as
continuous dynode electron multipliers for image intensifi-
cation applications (Kerr and Jorgensen, 1971).
Conventional electron multipliers used in photo multiplier
tubes have a discrete dynode structure, whereas a glass can
be fabricated into a continuous parallel plate or tube
dynode structure termed a channel plate multiplier.   A
voltage is applied to the ends of the plate to provide an
electric field for accelerating the electrons along the
device (see Fig.2).   Originally a lead silicate glass was
used with a reduced surface to enhance the operating
characteristics.   One of the improved semiconducting
glasses has a composition of B_2O_3(32.0%) Ce_2O_3(47.5%) Fe_2O_3
(20.5%) by wt. and a bulk room temperature resistivity of
10^9 ohm cm.   This type of material is advantageous in the
construction of a channel plate multiplier, consisting of
hundreds of fine tubes, in that no surface treatments are
required.

J. A. Savage

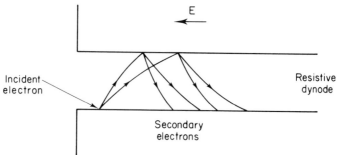

*Fig.2.   Electron multiplication
in a resistive glass dynode.*

B.   OPTICAL GLASSES

The advent of a demand for laser glass, and for glass
with very low absorption for fibre optic communications, has
raised severe materials problems.   For oxide lasers the
most important dope is neodymium oxide, and when dispersed
in a host medium it has an output signal at 1.06 μm.   The
optimum concentration in glass has been found to be about
5% by wt., from both practical and theoretical considerations.
The emission peak intensity, linewidth and fluorescent lifetime are dependent on the host glass composition as shown in
Table III (MacAvoy et al. 1965).

The best glasses for long lifetimes and high fluorescent efficiency are the silicates, and of these the alkali/
alkaline earth silicates are preferred.   The host glass
should possess good chemical durability and be capable of
being manufactured to a high standard of optical quality.
In practice the best compromise found for a commercially
manufactured composition has been a barium crown optical
glass or soda-lime-silica glass (Patek, 1970).   All impurities which have absorption bands overlapping the emission

bands of Nd^{3+} decrease the Nd^{3+} lifetime as well as absorbing
the signal.   This reduction in lifetime has been ascribed to
a resonance transfer mechanism.   The impurities which have
to be excluded are the transition elements such as Fe, Cu,
Ni, Co and the rare-earth elements Pr, Dy and Sm.   Nevertheless, with careful choice of raw materials and careful
processing it is possible to meet the technical requirements
for laser glasses.   The lifetime of Coming Code 0580 soda-
lime-silica laser glass is 320 μsec, and the linewidth is
324 $\overset{\circ}{A}$.   This is in contrast to a single crystal host such
as yttrium aluminium garnet, where the linewidth is of the
order of 6 $\overset{\circ}{A}$.   The applications for glass lasers are in
fields where the large linewidth is not a problem and where
large output powers are required.

TABLE III

Glass forming system	Emission peak μm	Emission peak, intensity, arbitrary scale	Lifetime μ sec	Linewidth $\overset{\circ}{A}$
Silicate	1.066	50-410	1000-150	240-480
Germanate	1.070	140	413-160	350
Borate	1.066	20-46	100-53	350
Phosphate	1.057	79-208	310-180	290
Tellurite	1.068	7500	159	280

With the increasing demand for communications it is
likely that the existing land lines and microwave links will
eventually become saturated.   An alternative approach would
be to use a laser beam to carry information and to use a
glass fibre as a light guide.   The use of a glass fibre
cable is attractive because it could replace some of the

present cable land lines in their existing conduits.   In
order to be economic, the booster or repeater stations in
this type of system should be not less than 1Km apart which
delineates the acceptable loss in the fibre at 20 dB/Km.
A few years ago a typical specification for the losses in
commercially available optical glasses was 20dB/m for low
index glass (N_d = 1.5) and 20dB/30cm for high index glass
(N_d = 1.8).   Although this loss is acceptable for the mass
market in optical components, it obviously does not meet the
requirement for optical communications.   From the earlier
discussion on phase separation it is clear that glass compositions have to be chosen carefully to avoid any phase
separation which might produce light scattering centres.
Extremely pure raw materials have also to be used to avoid
the transition and rare-earth element oxides which cause
absorptions in glass.   These two requirements can be met
but there remains a third, that of melting the raw materials,
homogenising the glass and drawing the fibres without re-
contamination from the atmosphere or furnace environment or
from the crucible used in the melting process.   Considerable
progress has been made in overcoming these problems and it
is possible to make bulk glass by a crucible melting process
which has an equivalent loss of 80-100 dB/Km.   Present
results indicate that it may be possible to meet the optical
communications requirement.

REFERENCES

Aranjo, R.J. (1968) Applied Optics 7, 781.
Bruckner, R. (1970) J. Non. Cryst. Solids 5, 123.
Bruckner, R. (1971) J. Non. Cryst. Solids 5, 217.

Drake, C.F., Scanlan, I.F. and Engel, A. (1969) Phys. Status Solidi 32.

Garfinkel, H.M. (1968) Applied Optics 7, 789.

Griffith, A.A. (1920) Phil. Trans. A221, 163.

Jones, G.O. (1956) Glass Methuen Monograph.

Kerr, J.I. and Jorgensen, L.L. (1971) J. Non. Cryst. Solids 5, 306.

MacAvoy, T.C., Charters, M.L. and Maurer, R.D. (1965) S.C.P. and Solid State Technology Feb. 23.

McMillan, P.W. (1964) Glass Ceramics, Academic Press.

Megla, G.K. (1966) Applied Optics 5, 945.

Morey, G.W. (1945) Glass Ind. 26, 417.

Mullin, J.B., Straughan, B.W. and Brickell, W.S. (1964) J. Phys. Chem. Solids 26, 782.

Patek, K. (1970) Glass Lasers, Iliffe Books.

Rawson, H. (1956) Proceedings IV International Congress on Glass, Paris, pp 62-69 Imprimerie Chaix, Paris.

Rawson, H. (1967) Inorganic Glass Forming Systems, Academic Press.

Stanworth, J.E. (1946) J. Soc. Glass Technol. 30, 54-64T.

Sun, K.H. (1946) "Glass Ind" 27, 552-554.

Turnbull, D. and Cohen, M.H. (1958) J. Chem. Phys. 29, 1049-1054.

Warren, B.E. (1937) J. Appl. Phys. 8, 645-654.

Weyl, W.A. (1959) Coloured Glasses, Dawsons, Pall Mall.

Zachariasen, W.H. (1932) J. Am. Chem. Soc. 54, 3841-3851.

OXIDE PHOSPHORS

M.J. TAYLOR

Royal Radar Establishment,
Malvern, Worcestershire

I.   INTRODUCTION

II.   CATHODOLUMINESCENT PHOSPHORS
    A.   GENERAL BACKGROUND
    B.   COMMENTS ON PARTICULAR CRT PHOSPHORS

III. FLUORESCENT LAMP PHOSPHORS
    A.   GENERAL BACKGROUND
    B.   COMMENTS ON PARTICULAR LAMP PHOSPHORS

I.   INTRODUCTION

A large number of oxide-dominated materials form useful phosphor systems, and applications include cathode-ray tube (CRT) screens, fluorescent lamps and scintillation counters. This article is primarily concerned with materials suitable for the first of these two applications.

The advent of the fluorescent lamp during the 1930's first stimulated serious development of oxide phosphors, and television has since provided a powerful incentive for CRT phosphor work. Sophisticated materials have now been developed for both these applications and the pace of research and development has consequently slackened. There is, however, a continuing effort to find better phosphors;

a small percentage increase in the efficiency of a lamp phosphor can give a substantial commercial advantage.

Phosphor technology is empirical. The reasons for this include the broad structureless nature of the absorption and emission bands of many systems and, until recently, the general unavailability of synthetic single crystals. The literature is extensive and no attempt is made in this article to be comprehensive. The aim has been to provide an introductory description of the most important materials, including some of the more recent developments.

A comprehensive survey of the luminescence of oxide-dominated lattices has been given by Johnson (1966). Other useful general references to luminescence in solids are a review by Garlick (1958) and books by Kroger (1948), Leverenz (1950), and Pringsheim (1949).

II.   CATHODOLUMINESCENT PHOSPHORS

Most of the important CRT phosphors fall into three categories, oxides, fluorides, and sulphides (selenides) of the ZnS type. Lists of phosphors and their characteristics are available from various manufacturers and from the American Joint Electron Devices Engineering Council (JEDEC 1966). Recent developments have been surveyed by Davis (1970). It is worth noting that the various phosphor classifications can be confusing, since they may refer either to the phosphor as a material or to the phosphor characteristics. For instance the JEDEC P-number refers to the phosphor characteristics, with a result that a given P-number may represent several completely different materials.

Some of the standard oxide phosphors are listed in

M. J. Taylor

TABLE I

Phosphor Material	Colour of Emission	Decay Time (to 10%)	JEDC Screen Classification	Uses
$Zn_2SiO_4:Mn$	YG; $\lambda_m = 530nm$	$2.45.10^{-2}s$	P1	Radar, oscillography.
$CaSiO_3:Pb,Mn$	O; $\lambda_m = 610nm$	$4.6.10^{-2}s$	P25	Radar, because of long persistence.
$(Zn,Be)SiO_4:Mn$ $(Ca,Mg)SiO_4:Ti$	W; $\lambda_m = 543$ & $610nm$ $\lambda_m = 427nm$	$1.3.10^{-2}s$ $5.5.10^{-5}s$	P18	Projection TV.
$Zn_2SiO_4:Mn,As$	G; $\lambda_m = 525nm$	$1.5.10^{-1}s$	P39	Integrating phosphor for low repetition rate displays and radar.
$Ca_2MgSi_2O_7:Ce$	BP; $\lambda_m = 335nm$	$1.2.10^{-7}s$	P16	Flying-spot scanner tubes, photography.
$Zn_3(PO_4)_2:Mn$	R; $\lambda_m = 640nm$	$2.7.10^{-2}s$	P27	Old standard red for colour TV.
ZnO	UV; $\lambda_m = 390nm$ G; $\lambda_m = 500nm$	$\lesssim 5.10^{-8}s$ $2.8.10^{-6}s$	P15	Flying-spot scanners, photography.
ZnO	G; $\lambda_m = 510nm$	$1.5.10^{-6}s$	P24	Flying-spot scanners.
$YVO_4:Eu$	OR; $\lambda_m = 618nm$	$9.10^{-3}s$	P22R	Colour TV.

R - red, O - orange, Y - yellow, G - green, B - blue, P - purple, W - white

Based largely on data in JEDEC publication 16A (1966) and in Davis (1970)

Table I, with brief comments on properties and uses, and several recently developed materials are listed in Table II. In this section the basic processes governing the operation of cathodoluminescent screens are outlined and a few specific systems are described in more detail.

TABLE II

MATERIAL	COMMENTS
Europium-doped materials, including YVO_4, Y_2O_3, Gd_2O_3, Y_2O_2S, La_2O_2S, Gd_2O_2S.	Efficient red emission, suitable for colour TV.
$SrSi_2O_5$:Pb, $BaSi_2O_5$:Pb, $Ca_3Si_2O_7$:Pb	Efficient UV emission, suitable for dry film photography.
$Y_3Al_5O_{12}$("YAG"):Ce	Efficient yellow-green emission with short persistence (<1 μs) suitable for flying-spot scanners.

A. GENERAL BACKGROUND

1. Origin of the luminescence

Although many materials exhibit an intrinsic cathodo-luminescence, most efficient phosphors are doped materials. With sulphide-type phosphors the emission has the characteristics of recombination radiation, and is usually associated with strong photoconductivity, suggesting that the impurities function as centres for electron-hole recombination. However, with many oxides and fluorides the emission is characteristic of the impurity and probably involves the initial formation of excitons which then migrate to, and excite, the activator centres. The mechanism by which the electron-beam energy is transferred to the phosphor system will be discussed in the next section.

The optimum impurity level reflects the luminescence mechanism; oxides and fluorides require an impurity level equal to about 1% of the number of cation sites, whereas sulphides require anything between one and three orders of magnitude lower. Mn is a common activator for oxides and Cu and Ag for sulphides. Other impurities can just as effectively act as 'killers' which quench the emission, and for this reason cleanliness is more important in the preparation of sulphide phosphors than for oxides and fluorides.

The luminescent decay characteristic often also reflects the nature of the luminescence mechanism. Generally speaking impurity-characterised emission from oxides and fluorides has an exponential time-decay whereas the recombination radiation from sulphides is more often non-exponential.

2. Basic considerations for CRT phosphors

The important phosphor parameters are (1) emission characteristics, such as efficiency, brightness, spectrum and build-up and decay time, (2) variation of luminescence output as a function of beam current and voltage, (3) resolution of the phosphor screen and (4) the chemical stability of the material during device processing and under electron bombardment. The widely differing requirements of various displays have led to the development of phosphors with a wide range of characteristics. For raster-type displays, the optimum persistence depends on the required refresh rate, which can be lower for data displays than for, say, TV. High definition flying-spot scanners require short persistence high efficiency emission matched to the spectral response of the photodetector. Long persistence

screens are required for radar, while fast oscillography
puts a premium on efficiency and fast rise time.   Phosphors
with an efficient ultraviolet output are needed for dry-film
photographic recording from the screen.

The cathodoluminescent efficiency of a given screen
depends on a number of factors in addition to the basic
efficiency of the luminescent process.   For most materials
something like 20% of the primary-beam energy is lost by
elastic back-scattering.   Primary electrons which are not
elastically back-scattered lose energy through inelastic
scattering and the production of secondary electrons.   Some
secondaries are also back-scattered and the ratio of secondaries emitted to primaries received depends on the screen-
to-cathode voltage;   for typical CRT voltages the ratio is
greater than unity and the screen charges positively.
Charging effects are usually eliminated in CRTs by coating
the screen with a thin layer of aluminium.   Since the
primaries penetrate to a depth which increases roughly as
the square of the accelerating voltage there is also a
minimum screen thickness for a given voltage below which
useful energy will be lost.

Those secondaries, the majority, which are not lost by
back-scattering, create further secondaries of lower energy,
and this process continues until the average energy of the
secondary electrons falls below the binding energy of deep
inner-shell electrons.   At this stage, energy loss to
valence-shell electrons, resulting principally in the formation of electron-hole pairs, becomes relatively more
important.   The mean energy ϵ for the production of
electron-hole pairs is related to the band-gap ϵ_g, but
exact calculations of ϵ have so far proved illusive;

estimates, and such data as is available, (Shockley, 1961; Klein, 1966; Kingsley and Ludwig, 1970) are consistent with $1.5\varepsilon_g < \varepsilon < 4\varepsilon_g$. Crude estimates of phosphor efficiencies can be made on the assumption that a primary electron of energy E produces E/ε electron-hole pairs, each of which yields one quantum of luminescence; the efficiency is then proportional to E_ν/ε, where E_ν is the energy of the luminescence transition. Estimates of this kind suggest that the highest measured efficiencies, in the region of 20% for materials such as ZnS:Ag, are near the theoretical maximum. For many other phosphors the efficiency is limited by losses in the transfer of energy between lattice and emitting centres, and by the low efficiency of the emission process itself (Ludwig and Kingsley, 1970). Silicate phosphors have efficiencies in the region of 5 to 10%. Although there are many shortcomings in the calculations outlined above, it is generally true that the most efficient phosphors are those with the smallest band-gap energies.

The dependence of luminescent output on electron-beam current density can be complex, varying from a pronounced saturating behaviour (sub-linearity) to a strong increase of fluorescence efficiency with current density (super-linearity). Superlinearity has not yet been observed in any oxide phosphor. In sulphides it is attributed to competition between activator and 'killer' impurity centres for the excitation energy. Saturation, on the other hand, becomes significant when the number of centres in the excited (emitting) state is comparable to the total number of active centres. For a power P in the primary electron beam, and a total of N activating centres the total luminescence intensity I is given by

$$I = P.N.E_\nu \left[\frac{n_a n_e}{1 + \tau n_a P} \right]$$

Where E_ν is the transition energy, τ is the luminescence
decay time, $n_a P$ is the probability per unit time that
activator centres are raised to the emitting level and n_e is
the probability that, once in this level, they decay
radiatively to the ground level.   It can be seen, therefore,
that long persistence is associated with a low saturation
threshold and consequent low brightness.   This is particularly evident for fluorides, which provide some of the
longest available decay times ($\sim$ seconds) and are thus
widely used in radar applications.   It also seems likely
that the low activator levels of sulphides are the reason
why these phosphors saturate much more easily than oxides
(Meyer and Palilla, 1969), leading to the greater use of
oxides in high brightness screens.

Decay times are determined by the strength of the
transition and by the efficiency, n_e, of emission from the
excited state.   Very short decay times can result from a
high radiative transition probability or from efficient
quenching processes, but in the latter case the efficiency
n_e will be low.   Decay times of $\lesssim$ 10 nsecs occur in the
short wavelength emission of some oxides (para IIB2) and
also in the visible edge emission of CdS and CdSe.   There
is usually a direct relationship between rise-time and decay
time, the slower the decay the slower the rise.

Although all phosphors must saturate at some level of
current density, in practice a lower limit to the attainable
brightness is often encountered at high beam current levels

due to screen heating and material damage. Many phosphors
suffer from fatigue and discolouration ('dark-burn') under
these circumstances, but although the study of radiation
damage has in general become a vast subject, comparatively
little attention has been given to damage mechanisms in
luminescent solids. Damage in CRT screens is commonly
attributed to bombardment by residual negative ions, since
these can transfer a greater fraction of their momentum to
the lattice than can electrons of the same energy. Zinc
silicate is about ten times as resistant to damage by ion
bombardment as zinc sulphide and indeed high stability is a
general feature of oxide-dominated lattices.

The mechanism of damage may involve the displacement
of lattice anions; Grosso et al. (1967) have correlated
fatigue in various CRT screens with the accumulation of
characteristic vapours in the tubes, oxygen for oxide phosphors, sulphur for sulphides and fluorine for fluorides.
Dark-burn is sometimes due to colour centre creation and is
particularly serious with fluorides.

The resolution of a CRT screen is primarily limited by
the particle size. Recent progress in synthesis techniques
has cast some doubt on the earlier idea that phosphor
efficiency increases with particle size, and there have been
reports (Davis, 1970) that phosphors with an average particle
size of 0.5μm can now be synthesised.

B.   COMMENTS ON PARTICULAR OXIDE CRT PHOSPHORS

1. Silicates

Table I shows that silicates form the biggest group of
oxide phosphors. Of these, Zn_2SiO_4:Mn (rhombohedral
willemite structure) is the most well known. The

characteristic green emission from Zn_2SiO_4:Mn is a standard
for relative measurements of luminescent efficiency.
Willemite-type phosphors are often used in oscilloscopes
because their characteristics offer a good compromise between
the requirements of high and low speed scanning.

A typical method of synthesising Zn_2SiO_4:Mn consists
of firing a mixture of ZnO and SiO_2 at $1000^\circ C$ to $1450^\circ C$,
together with a small quantity of a manganese salt and a
chloride catalyst; optimum manganese concentration for the
green emission is between 1 and 5%. The exact heat treatment is critical and under certain conditions it is possible
to shift the broad emission band to yellow and red wavelengths and also to get narrow-banded emission. Osiko and
Maksimova (1960) have recently confirmed, by chemical
analysis, Kroger's original suggestion (Kroger, 1948), that
the broad-banded emission is due to Mn^{2+} and the narrow-
banded emission to Mn^{4+}. Current work (Johnson, 1966)
suggests that the various Mn^{2+} bands are due to Mn^{2+} on
different sites; the green band due to Mn^{2+} on Zn sites,
and tetrahedrally co-ordinated by four oxygens; and the
longer wavelength bands due to Mn^{2+} at interstitial sites,
where it is octahedrally co-ordinated by six anion nearest
neighbours. The primary effect of adding arsenic to
Zn_2SiO_4:Mn(P39) is to enhance the decay time, which becomes
non-exponential. A survey of Mn activated phosphors by
Fonda (1957) has shown that there is a correlation between
the Mn^{2+} emission energy and the proximity of its nearest
neighbours, a more crowded environment leading to a higher
emission energy. Theoretical studies of Mn^{2+} activated
oxide phosphors have been made by Orgel (1955) and Butler
(1966).

Zn_2SiO_4:Mn CRT screens sometimes exhibit an increase
of efficiency during the first few hours of use.   This may
be due to the action of the electron beam in converting
manganese in higher valence states to Mn^{2+}.

Zinc beryllium silicate (P18) is not much used now.
White emission can be obtained by replacing it with a
mixture of P1 and Eu red (see section 3).

Many silicates activated by Pb^{2+} give efficient blue
or ultraviolet emission (see also section 5).   The addition
of Pb to Mn-activated calcium silicate (P25) perturbs the
Mn^{2+} centres sufficiently to shift the emission colour from
yellow to orange, and to substantially increase the decay
time.

2. Zinc oxide

Zinc oxide differs from most other oxide phosphors in
being a semiconductor, and provides an excellent example of
the progress made by recent research towards understanding
the luminescent mechanisms of well-known phosphors.   Zinc
oxide phosphors are widely used in CRT applications which
require a rapid decay, but it was only recently demonstrated
that the luminescence is not, as had previously been widely
accepted, an intrinsic phenomenon, but depends on trace
impurities such as Cu.

Zinc oxide crystallises as hexagonal rods with the
wurtzite structure.   The phosphor is commonly synthesised
by burning zinc in air, but the exact treatment is a critical
factor in determining the nature of the luminescence.   The
emission spectrum consists of a narrow band in the UV and a
broad band in the green spectrum, and the ratio of these
depends on the method of preparation (Heiland et al. 1959).

The UV band is enhanced when zinc oxide is prepared by
firing in an oxidising atmosphere, whereas the green band
is enhanced by firing at high temperatures ($1200^{\circ}C$) in a
reducing atmosphere such as hydrogen or carbon monoxide.
It is thought that the latter treatment generates excess
zinc.   There is a clear correlation between electrical
conductivity and the intensity of green luminescence;   the
luminescence is more intense the smaller the conductivity.
The decay time of the green emission also varies with the
method of preparation, and a more rapid decay is associated
with a lower efficiency.

Dingle (1969) has recently produced convincing evidence that the green emission is due to traces of copper.
In his crystals the level of Cu was $\sim$ 4ppm.   The broad-
band emission is accompanied at low temperatures by a sharp
zero-phonon line at 2.8590 eV, and Dingle was able to
correlate g values in the Zeeman pattern of this line with g
values of Cu^{2+} on zinc sites obtained by other workers from
the ESR of doped material.   He suggests the transition is
due to charge transfer between a highly shielded, localised
level of Cu^{2+} and a level which is strongly perturbed by the
valence-band states of the crystal.   Traces of other
impurities, such as Mn can reduce the efficiency of luminescence from ZnO.

3. Europium-activated materials

A considerable improvement of the red output of colour
TV screens has recently been achieved by replacing
$Zn_3(PO_4)_2$:Mn with Eu-activated materials;   previously,
inconveniently high electron-beam currents had been necessary
to compensate for the low luminous efficiency of the

phosphate.   Very efficient europium emission can be obtained
from a number of Eu-activated host lattices, including
yttrium oxide, yttrium vanadate, and yttrium, lanthanum and
gadolinium oxysulphides.   The vanadate, YVO_4:Eu, was the
first to be adopted for colour TV (Levine and Palilla, 1965;
Burrus and Paulusz, 1969) but yttrium oxysulphide, Y_2O_2S:Eu
subsequently turned out to be more efficient (Yocom and
Schrader, 1968;   Hardy, 1968) and is now used in the RCA
system.   However, the oxysulphide suffers more from burn
and colour shift at high electron-beam intensities.

The emission from europium-activated phosphors consists of a number of narrow-lines characteristic of Eu^{3+},
and due to transitions between states of the shielded inner
4f electrons.   Electric dipole transitions between states
of the same electron configuration are parity forbidden, and
the transitions only occur with high probability due to the
action of crystal fields of relatively low symmetry in
admixing the 4f states with states of different electron
configuration.   However, due to the shielding effect of the
outer 5s and 5p electrons the crystal field interaction is
much less than for transitions of iron group elements, and
is not sufficient to cause much broadening of the 4f levels.
The lines thus remain sharp and are not very sensitive to
temperature variations;   the 611.3 nm line from Y_2O_3:Eu
retains its efficiency up to 650°C.

The distribution of fluorescent energy from YVO_4:Eu
and Y_2O_3:Eu is shown in Fig.1a.   In both cases the dominant
emission is due to transitions which originate from the $^5D_0$
level and terminate on states belonging to the $^7F_2$ multiplet.
In the case of YVO_4:Eu about 70% of the total fluorescent
energy is contained in two lines, at 619 nm and 615 nm, both

due to $^5D_0$-$^7F_2$ transitions. The remaining fluorescent
energy, in the case of YVO$_4$:Eu, is distributed mainly
between lines at 698 nm due to $^5D_0$-$^7F_4$ transitions and lines
in the vicinity of 595 nm due to $^5D_0$-$^7F_1$ transitions. The
lines at 619 and 615 nm are less than 1 nm in width, the
character of the emission being independent of the mode of
excitation, whether cathode rays or ultraviolet radiation.

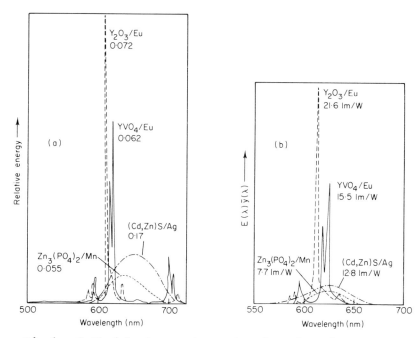

*Fig. 1. Cathodoluminescent spectra from four important
red-emitting phosphors, (a) in terms of the spectral
energy distribution, and (b) in terms of the luminosity
distribution, where E(λ) is the radiant energy
distribution and $\overline{y}$(λ) is the eye response function.
Energy efficiencies are given next to each trace in (a)
and luminous efficiencies in (b).*
(After Levine and Palilla, 1965).

Fig.1a also shows the output from $Zn_3(PO_4)_2$:Mn, one of the first red TV phosphors, and the relative luminosities of these three phosphors are compared in Fig.1b.   Although Y_2O_3:Eu is the more efficient, YVO_4:Eu was selected for colour television because its emission has a deeper red colour.   The emission from Y_2O_2S:Eu has a dominant doublet which is shifted from 3 nm to the blue from the YVO_4:Eu doublet and has some 30% greater efficiency, depending on the electron-beam loading.   The introduction of the oxy-sulphide has largely eliminated the red deficiency of colour TV screens.

4.   Cerium-activated phosphors

$Ca_2MgSi_2O_7$:Ce (P16) is one of the more efficient of a number of cerium activated materials that give strong, short-persistence luminescence in the blue and ultraviolet spectral regions (Gilliland and Hall, 1966).   New materials of this type continue to be reported and one of particular interest (see Table II) is $Y_3Al_5O_{12}$:Ce, which has the garnet structure and gives strong visible emission (Holloway and Kestigan, 1967;   Blasse and Bril, 1967).

In most cases the emission consists of broad, struc-tureless bands and this, together with the evident sensitivity to the crystalline environment, complicates the assignment problem.   It seems likely that Ce^{3+} ions are responsible for the emission from most of these phosphors (Blasse and Bril, 1967), although in the case of $Y_3Al_5O_{12}$:Ce it has also been suggested that cerium ions of higher valency or even cerium ion-pairs could be involved (Holloway and Kestigan, 1967).

Ce^{3+} has a single 4f electron but unlike other rare

earths, e.g. Eu, the emission cannot be correlated with
intraconfigurational transitions. It has been proposed
that 4f-5d transitions are involved, which would account for
the strength and short lifetime of the emission since these
are allowed electric-dipole transitions. Also, since the
5d electrons are not as efficiently shielded as the 4f
electrons, strong environmental effects are more easily
accounted for. Blasse and Bril suggest that the unusually
long-wavelength visible emission from $Y_3Al_5O_{12}$:Ce comes from
Ce^{3+} ions on yttrium sites; the low symmetry (distorted
cubic) of these sites gives a large crystal field splitting
of the appropriate 5d level and emission originates from the
lowest component.

$Y_3Ga_5O_{12}$:Ce does not show any cathodoluminescence,
which may be the result of inefficient transfer of energy
from the lattice to the activator. However, partial
substitution of Ga in the Al garnet shifts the emission to
shorter wavelengths, and the material then also exhibits
some phosphorescence.

5. UV-emitting phosphors

A range of new oxide phosphors has been developed over
the last few years with high UV output which are useful for
photographic work. Many contain lead as an activator and
most are silicates. The spectral output of those included
in Table II is shown in Fig.2. Decay-times have been
measured by Sisneros (1967).

III. FLUORESCENT LAMP PHOSPHORS

It was the advent of the fluorescent lamp in the
1930's that greatly stimulated the technical development of

oxide dominated phosphors.   Today such lamps are used
widely for lighting and also in specialised applications
such as photoprinting.

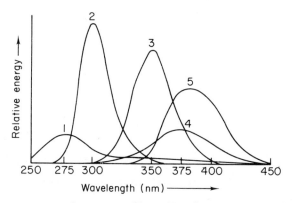

*Fig.2.   Spectral energy distribution for UV-emitting
phosphors.*

   (1) SrSi$_2$O$_5$:Pb   (2) ITT phosphor JJ-13K

   (3) BaSi$_2$O$_5$:Pb   (4) Ca$_3$Si$_2$O$_7$:Pb   (5) P-16

(After Davis, 1970).

A.   GENERAL BACKGROUND

     In the fluorescent lamp an electrical discharge in
mercury vapour gives rise to strong UV emission which in
turn excites broad-band visible emission from a coating of
phosphor on the wall of the discharge tube.   There are two
widely used types of lamp, the familiar low pressure discharge lamp and the high pressure mercury vapour (hpmv) lamp.
Their UV emissions are predominantly at 254 nm and 365 nm
respectively, and it is necessary that the phosphor should
have excitation characteristics matched to the emission in a
particular lamp.   In some phosphors UV energy is absorbed
directly by the emitting centres, in others, UV energy is

absorbed by the lattice or by a 'sensitising' impurity and then transferred to the emitting centres.

In contrast to the low pressure lamp the hpmv lamp has a considerable visible output in the blue-green, so that the prime consideration for the hpmv lamp is to provide a phosphor with an orange or red emission to correct for the deficiency of these colours.

The requirements for lamp phosphors are much simpler than for CRT phosphors; the important parameters are (1) efficiency, (2) the emission spectrum, (3) the material stability and (4) for hpmv lamps, the change of emission characteristics with temperature. Most modern lamp phosphors are oxides because they are more efficiently excited by the mercury UV radiation than other materials; sulphides are only very occasionally used nowadays for colour correcting purposes.

The earliest lamps used $Zn_2SiO_4:Mn^{2+}$ (see Fig.3), which has been described in the section on CRT phosphors. The requirement for white emission then led to the use of a mixture of phosphors, Mn-doped cadmium borate (red), $Zn_2SiO_4:Mn$ (green) and $CaWO_4$ (blue). However the 'whiteness' of lamps made with this mixture tended to vary due to the difficulty of maintaining fixed ratios of the component concentrations. From 1939 to 1950 mixtures of $(Zn,Be)_2SiO_4:$ Mn^{2+} (Table I) and $MgWO_4$ were used in 'Daylight' and 'Warm-white' lamps. However the toxicity of the silicate (due to beryllium) and the prospect of reduced materials manufacturing costs led to the choice of halophosphates, $Ca_5(PO_4)_3(Cl,F):$ Sb^{3+},Mn^{2+}. In these doubly activated materials the Sb^{3+} and Mn^{2+} give two overlapping bands in the visible region and good colour control is possible by varying the relative

TABLE III

MATRIX	ACTIVATORS	COLOUR OF EMISSION	COMMENTS
Calcium Tungstate, $CaWO_4$	- Pb	Deep Blue Pale Blue	Mainly in blue lamps.
Barium Disilicate, $BaSi_2O_5$	Pb	UV peak at 350nm	For long UV emission.
Zinc Orthosilicate, Zn_2SiO_4	Mn	Green	Mainly in green lamps.
Calcium Metasilicate, $CaSiO_3$	Pb, Mn	Yellow to Orange	In 'de luxe' colour lamps.
Cadmium Borate, $Cd_2B_2O_5$	Mn	Orange-Red	Mainly in red lamps.
Barium Pyrophosphate	Ti	Blue-White	
Strontium Pyrophosphate	Sn	Blue	
Calcium Halophosphate, $Ca_5(PO_4)_3(Cl,F)$	Sb, Mn	Blue to Orange and White	Main group of lamp phosphors. Also strontium halophosphates.
Strontium Orthophosphate (containing Zn or Mg), $(Sr, Zn)_3(PO_4)_2$	Sn	Orange	In 'de luxe' colour lamps, and also high efficiency, high pressure lamps for colour correction.
Magnesium Arsenate, $Mg_6As_2O_{11}$	Mn	Red	In 'de luxe' colour lamps. Emission attributed to Mn^{4+}.
Magnesium Fluogermanate $3MgO.MgF_2.GeO_2$	Mn	Red	Comment as for strontium orthophosphate. Emission attributed to Mn^{4+}.
Yttrium Vanadate, YVO_4	Eu	Red	Colour correction in high pressure lamps.
Magnesium Gallate $MgGa_2O_4$ (Aluminium substituted)	Mn	Green	Photoprinting.

concentration of Sb and Mn and also the (Cl/F) ratio.
Because of their commercial importance these materials are
now produced in large quantities; by 1966, for instance
there was an annual production of 3.10^6 kg of the halophosphates.

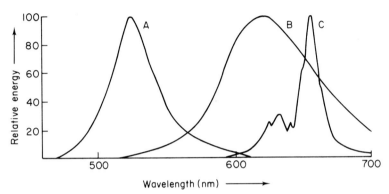

*Fig.3.   Spectral energy curves of manganese-activated
phosphors: (A) $Zn_2SiO_4:Mn^{2+}$ (B) $Cd_2B_2O_5:Mn^{2+}$
(C) $3MgO.MgF_2.GeO_2:Mn^{4+}$*

(After McKeag, 1965).

Recent developments have mainly concerned the detail
of lamp performance, the optimisation of efficiency, colour
correction and improved temperature stability. The effect
of temperature is important in high pressure lamps in which
the phosphor may be required to operate at $\sim$ 250°C. At
these temperatures many phosphors show a marked diminution
in brightness, with a spectral broadening of the emission
and a shift to shorter wavelengths. Thus, since they must
also be excited by 365 nm radiation, the choice of phosphors
for high pressure lamps is very limited. Of the phosphors
shown in Table III as being suitable for this purpose, the
fluorogermanate is almost ideal but is expensive to

manufacture because of the germanium content.   The recently developed Eu activated phosphors show great potential in this field; $YVO_4:Eu^{3+}$ gives about 10% greater output than magnesium germanate, gives improved colour and has excellent temperature stability.

The stability of the luminescence output with time, or 'lumen maintenance', has been the subject of many investigations.   There is usually a fairly rapid decrease in output of a few percent in the first few hours operation, followed by a slower rate of decline amounting to some 15% or so towards the end of the lamp life.   The latter is probably due to the deposition of mercury from the discharge onto the phosphor, and consequent reactions with the surface, but the former is thought to be associated, at least in the halophosphates, with the formation of colour centres.

The quantum efficiencies of most lamp phosphors are high, in the region of 0.7 to 0.9 (Tregellas-Williams, 1958; Johnson, 1961).   However the luminous efficiency depends on colour, and the improvement of colour-rendering by the addition of red-emitting phosphors therefore leads to lower luminous efficiency.   This has led to the development of two types of lamp to meet commercial requirements, one based mainly on the halophosphates, giving high efficiency standard colours at around 70 lm/W, and the 'de-luxe' range with considerably improved colour rendering properties but some 30% less efficiency.   Table III includes some of the phosphors which are used for enhancing the red output in 'de-luxe' colours.

B.   COMMENTS ON PARTICULAR LAMP PHOSPHORS

1. Halophosphates

A precis of the early literature on halophosphate phosphors has been given by Johnson (1961).   Halophosphates
have the apatite structure and tolerate extensive substitution for Ca, P and the halogen.   The luminescence of
material which is doubly doped with manganese and antimony
consists of two broad bands, peaking at about 565 nm and
470 nm, which have been attributed, largely from chemical
evidence, to Mn^{2+} and Sb^{3+} respectively.   The Mn^{2+} emission
is indeed very similar to the emission from Zn_2SiO_4:Mn.

Recent progress in growing large single-crystals of
these materials has led to a more sound understanding of the
manganese emission.   Manganese substitutes for calcium on
the two main calcium sites in halophosphates, termed Ca(I)
and Ca(II).   Ca(I) is co-ordinated by six nearest-neighbour
oxygens, in the form of a slightly twisted triangular prism
(C_3 point symmetry) while Ca^{2+} ions on Ca(II) sites sit at
the corners of equilateral triangles with an F^- ion at the
centre (C_{1h} point symmetry).   The distribution of manganese
between these sites depends on the halide ratio (Johnson,
1961) and the heat treatment during synthesis (Apple and
Ishler, 1962).   Ryan et al. (1970) have been able to
identify the excitation and emission spectra of Mn^{2+} on both
of these sites by means of a detailed correlation of the
strength of these spectra and ESR spectra, as the distribution of Mn^{2+} amongst the two sites was varied.   Both
manganese emissions peak in the green spectral region, with
a mutual shift of only about 10 nm, but they differ in
polarisation.   In most of the crystals used by Ryan et al.
Mn(I) emission was dominant but the proportion of Mn(II)

centres increased strongly as the manganese concentration was increased above 1%.   Other emissions, seen at low temperatures, were attributed to Mn(II) centres having nearby oxygen vacancy-complexes.

The luminescence due to antimony is not well understood, although it has been suggested that charge-transfer transitions of complexes such as $Sb^{3+}-O^{2-}$ may be involved (Williams, 1960).

A typical method of synthesis consists in mixing calcium hydrogen phosphate with the chloride or fluoride of calcium, strontium or ammonium, together with the necessary activating impurities, and firing the mixture in an inert atmosphere at 1050° to $1200^{\circ}C$.   Chloride evaporation can be a problem.   It is also important to avoid conditions that lead to oxidation of Mn^{2+} to Mn^{3+} and Sb^{3+} to Sb^{+} (Wanmaker, 1956).

Many impurities, particularly the transition elements, other than Mn, reduce the luminescent efficiency, but others, including Ce and Pb actually enhance the brightness by a small factor.   Calcium and strontium halophosphate phosphors activated by Ce and Mn can be as efficient as those activated with Sb and Mn.

The loss of efficiency of halophosphate phosphors during the first few minutes of lamp operation has been correlated with absorption by colour centres which are formed by the action of 184.9 nm resonance radiation from the mercury discharge (Johnson, 1961; Apple and Aicher, 1966).   Absorption peaks are generated at 370, 450 and 610 nm (Apple, 1963).   The addition of certain elements, especially Cd, substantially reduces colour centre generation and also the loss of brightness.   More information

about the nature of these colour centres has resulted from
work with large single crystals of chloro- and fluoro-
phosphates (Swank, 1964; Piper, Krautz and Swank, 1965).
The primary centres appear to be electrons trapped at
halogen vacancy sites (F-centres) but there is also evidence
for the existence of M-centres.

2. YVO_4:Eu and Y_2O_3:Eu phosphors

Red-emitting phosphors have been described in some
detail in the section on CRT phosphors.   The only other
consideration in the context of lamp phosphors is the
excitation spectrum, the emission spectrum being independent
of the mode of excitation. YVO_4:Eu is much more efficiently
excited by UV radiation than Y_2O_3:Eu.   Fig.4 compares the
excitation spectra of the two systems.   The intense, long
wavelength, excitation band of YVO_4:Eu is due to energy
absorption by the host lattice followed by transfer to the
Eu^{3+} ions.   Fluorescence due to direct excitation of Eu^{3+}
ions is weak compared to that due to excitation in the host
band.

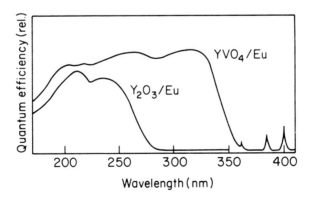

Fig.4. Excitation Spectra of YVO_4:Eu and Y_2O_3:Eu.
(After Levine and Palilla, 1965).

3. Magnesium gallate phosphors

Manganese-activated magnesium gallate (spinel structure) has an efficiency comparable to that of Zn_2SiO_4:Mn (Wanmaker, ter Vrugt and de Bres, 1967), and the emission is also probably due to Mn^{2+}. Incorporation of aluminium shifts the excitation bands to shorter wavelengths and raises the quenching temperature.

REFERENCES

Apple, E.F. and Aicher, J.O. (1966) Proc. Internat. Conf. on Luminescence (pub. 1968, G. Szigeti ed.) Akad Kiado, Budapest, Hungary, pp 2013-26.

Apple, E.F. (1963) J. Electrochem. Soc. 110, 374.

Apple, E.F. and Ishler, W.E. (1962) "Luminescence of Organic and Inorganic Solids", (H.P. Kallman and G.M. Spruch eds.) pp 576-595, Wiley, New York.

Blasse, G. and Bril, A. (1967) J. Chem. Phys. 47, 5139-5145.

Butler, K.H. (1966) Proc. Internat. Conf. on Luminescence (pub. 1968, G. Szigeti ed.) Akad Kiado, Budapest, Hungary, pp 1313-29.

Burrus, H.I. and Paulusz, A.G. (1969) J. Sci. and Technol. 36, 105-113.

Davis, J.A. (1970) "Recent Advances in Display Media", Symposium Proceedings, pp 25-39.   Technol. Pub. Corp. Los Angeles, Calif.

Dingle, R. (1969) Phys. Rev. Letts. 23, 579-81.

Fonda, G.R. (1957) J. Opt. Soc. Am. 47, 877-880.

Garlick, G.F.J. (1958) "Luminescence", Handbuck der Phys. XXVI, 1-128.

Gilliland, J.W. and Hall, M.S. (1966) Electrochem. Tech. 4, 378.

Grosso, P.F., Taylor, R.C. and Ward, S.A. (1967) J. Appl.
Phys. 38, 2697-8.

Hardy, A.E. (1968) I.E.E.E. Trans. Elect. Dev. ED-15,
868-872.

Harrison, D.E. and Hoffman, M.V. (1959) J. Electrochem. Soc.
106, 800-804.

Heiland, G., Mollwo, E. and Stuckmann, F. (1959) Sol. St.
Phys. 8, 191-323.

Holloway, W.W. and Kestigan, M. (1967) Phys. Letts. 25A,
614-15.

Johnson, P.D. (1961) J. Opt. Soc. Am. 51, 1235-38.

Johnson, P.D. (1966) "Luminescence of Inorganic Solids"
(P. Goldberg ed.) pp 287-336, Academic Press, New York.

JEDEC (1966) "Optical Characteristics of CRT Screens"
publication number 16A.

Kingsley, J.D. and Ludwig, G.W. (1970) J. Electrochem. Soc.
117, 353-359.

Klein, C.A. (1966) J. Phys. Soc. Japan 21, suppl.307.

Klick, C.C. (1955) Brit. J. Appl. Phys. suppl. 4, 574-578.

Kroger, F.A. (1948) "Some Aspects of Luminescence of Solids"
Elsevier, Amsterdam.

Leverenz, H.W. (1950) "Luminescence of Solids", Wiley,
New York and Chapman and Hall, London.

Levine, A.K. and Palilla, F.C. (1965) Int. Symp. on
Luminescence, pp 317-324, Verlag Karl Thiemig K.G.,
Munchen.

Ludwig, G.W. and Kingsley, J.D. (1970) J. Electrochem. Soc.
117, 348-353.

McKeag, A.H. (1965) GEC Journal 32, 21-28.

Meyer, V.D. and Palilla, F.C. (1969) J. Electrochem. Soc.
116, 535-539.

Orgel, L.E. (1955) J. Chem. Phys. 23, 1958.

Osiko, V.V. and Maksimova, G.A. (1960) Opt. Spectry. (USSR), English Trans. 9, 248-9.

Piper, W.W., Kravitz, L.C. and Swank, R.K. (1965) Phys. Rev. 138A, 1802-1814.

Pringsheim, P. (1949) "Fluorescence and Phosphorescence", Interscience, New York.

Ryan, F.M., Ohlmann, R.C., Murphy, J., Mazelsky, R., Wagner, G.R. and Warren, R.W. (1970) Phys. Rev. (B) 2, 2341-52.

Schulman, J.H. (1946) J. Appl. Phys. 17, 902-908.

Sisneros, T.E. (1967) Appl. Optics 6, 417-420.

Schockley, W. (1961) Solid State Electronics 2, 35.

Swank, R.K. (1964) Phys. Rev. 135A, 266-275.

Tregellas-Williams, J. (1958) J. Electrochem. Soc. 105, 173-178.

Wanmaker, W.L. (1956) J. Phys. Rad. 17, 636-640.

Wanmaker, W.L., ter Vrugt, J.W. and de Bres, J.G.C.M. (1967) Philips Res. Repts. 22, 304-308.

Williams, F.E. (1960) Electrochem. Soc. Abstr. 9(49), 40-42.

Yocom, P.N. and Schrader, R.E. (1968) Proc. 7th Rare-Earth Research Conference, Coronado, Calif. p 601.

FERROELECTRIC CERAMIC OXIDES

F.W. AINGER

The Plessey Company Limited,
Towcester, Northamptonshire

I.   INTRODUCTION

Ferroelectricity was first discovered fifty years ago
in Rochelle salt by Valasek (1921), and has been associated

mainly with single crystals.   A ferroelectric crystal is
defined as a crystal which exhibits a spontaneous polarisation, the direction of which may be reversed by an electric
field.   Consequently all ferroelectrics are piezoelectric
(i.e. change polarisation with stress) and pyroelectric (i.e.
change polarisation with temperature) whilst the converse is
not true.   The distinguishing feature of a ferroelectric
crystal is that the polarisation direction coincides with
either the positive or negative direction of one of the
crystallographic axes.   The components of polarisation are
only opposite in the sense of polarity and are contained
within an array of identical unit cells which may be put
together without introducing strain.   These regions of
single polarity are termed domains and negatively and positively polarised domains are related in the same way as a
twin.   However, their behaviour is different from the twins
usually found in crystals since they can be very mobile and
can be moved by both electrical and mechanical stresses.

Two further significant properties associated with
ferroelectricity are dielectric hysteresis and high permittivity.   The switching of ions from one stable position to
another requires energy, and the relationship between the
electric displacement (polarisation) and electric field
exhibits hysteresis if sufficiently large fields are applied
to effect the switching.   If a crystal contains equal
numbers of positive and negative domains the overall polarisation is zero.   When a field is applied in the positive
direction a linear relationship (OS in Fig.1) is observed
between polarisation (P) or electric displacement (D) and
the applied field (E) for small values of E.   As the field
is increased, more negative domains switch over to the

opposite direction, and the polarisation rises rapidly until
a saturation value is reached at T, when the crystal is
single domain. On reversing the field, the domains do not
readily switch back to the original state but remain aligned
to give a remanent polarisation OU at zero field. In order
to destroy the overall polarisation, a field equivalent to
OV, the coercive field, is required. At slightly higher
fields the crystal saturates in the opposite sense, at W.
Ferroelectric hysteresis loops may be displayed using the
circuit originally reported by Sawyer and Tower (1930).

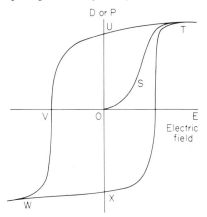

Fig. 1. Hysteresis loop for ferroelectrics.

Much of the original work on ferroelectricity was
carried out with crystals which were fairly readily grown
from aqueous solution. These crystals include dihydrogen
phosphates and arsenates (of which the ammonium and potassium
salts, $NH_4H_2PO_4$ (ADP) and KH_2PO_4 (KDP) are the best known),
tartrates, sulphates and nitrates. Although these water
soluble crystals are fragile and require particular care in
their cutting and polishing, they have been used for a range
of piezoelectric and pyroelectric devices. Crystals of
comparatively large dimension are often required, and the

natural growth habit of the single crystal does not always
lend itself readily to a specific crystallographic cut.
Minor impurities such as metal ions added to the solvent may
act as habit modifiers to give a more suitable crystal form,
provided that they do not destroy the ferroelectric behaviour
of the required crystal.   About twenty-five years ago it was
found that certain double oxide compounds possessed large
enough dielectric constants to be potentially useful for the
design of capacitors with relatively high values of
capacitance per unit volume.   De Bretteville, Jr. (1946)
found that barium titanate ceramics exhibited saturated
hysteresis loops and thus confirmed the conclusions of Wul
and Vereshchagin (1945) that barium titanate was a ferroelectric.

        Modern ceramic materials embrace a wide range of
inorganic materials in which oxides are pre-eminent.
Ceramics are made from unreacted, partially reacted, or
fully reacted powders which are sintered to a low or zero
porosity state by well defined heat treatments.   Ceramic
products are brittle, and final machining to a specific
shape can be expensive.   However, an advantage of ceramic
processing is that a ceramic can be shaped during one of the
final fabrication stages provided shrinkage during sintering
is taken into account.   The combination of a traditional
ceramic technology and modern ceramic material has produced
rapid development of new electronic materials, many of which
are based on ferroelectric systems.

II.   FERROELECTRIC OXIDE SYSTEMS

        The various types of ferroelectrics discussed are
limited to the oxygen octahedral ferroelectrics consisting

of one or more component oxides which have been fabricated
and used in the polycrystalline form. A common feature of
all these materials is the BO_6 octahedral building block,
although the materials may have different crystal structures,
electrical and mechanical properties, Curie temperatures,
and polarisations. The fundamental nature of the BO_6
octahedron and its relative importance to the electro-optic
and non-linear optical properties of ferroelectrics has been
amply covered by Wemple and DiDomenico (1968, 1969).
Classification of ferroelectrics is sometimes based on the
number of directions allowed for the spontaneous polarisation,
which is useful for domain formation studies. Additional
classifications are based on the type of phase change
occurring at the Curie temperature T_c (defined as the temperature at which the polar to non-polar transition occurs),
which is displacive in many oxide systems, and on whether or
not the non-polar phase exhibits a centre of symmetry in its
point group. Here, ferroelectrics are classified according
to their crystal structure, since it is intimately related to
the allowable polarisation directions and domain structure.
The pyrochlore, tungsten bronze and more complex oxide
crystal types are briefly discussed whilst the perovskite
group are described in detail using selected members to
illustrate ceramic processing and applications.

A.   PYROCHLORE TYPE FERROELECTRICS

        This group has a general formula of $A_2B_2O_7$ with struc-
tures similar to the mineral pyrochlore, $CaNaNb_2O_6F$. These
pyrochlores can have the formulae $A_2^{2+}B_2^{5+}O_7$ or $A_2^{3+}B_2^{4+}O_7$
according to the valency of the A and B cations. The best
known ferroelectric member of this family is cadmium

pyroniobate $Cd_2Nb_2O_7$ which is cubic at room temperature.
Below the Curie temperature of 185°K it assumes a slightly
distorted cubic form which is probably tetragonal with a c/a
ratio of 1.0005.   Dielectric measurements on the ceramic
form were made by Hulm (1953) who found evidence for a
further transition at 85°K.   No commercial exploitation of
this material, or of the solid solutions it forms with
$Ca_2Nb_2O_7$, $Cd_2Ta_2O_7$ or $Pb_2Nb_2O_7$, which have lower values of
T_c than pure $Cd_2Nb_2O_7$, has been reported.

B.   TUNGSTEN BRONZE FERROELECTRICS

    A revival of interest in ferroelectrics having the
tungsten bronze type structure has recently taken place
because of their potential use as single crystals in non-
linear optics.

    The tetragonal tungsten bronze unit cell (after
Jamieson et al. 1968) is shown in Fig.2 in the projection of
the (001) plane and can accommodate metal ions in five
different sites designated A_1, A_2, C, B_1 and B_2.   The
structure consists of a complex array of distorted BO_6
octahedra sharing corners in such a way that there are three
different types of interstices available for cation occupa-
tion.   The A_1 sites are in the pentagonal tunnels parallel
to the c axis of which there are four per unit cell, each
surrounded by fifteen oxygen atoms.   The A_2 sites are in
the tetragonal tunnels parallel to the c axis of which there
are two per unit cell, each surrounded by twelve oxygen
atoms.   The C sites are in the trigonal channels parallel
to the c axis of which there are four per unit cell, each
surrounded by nine oxygen atoms.   The B_2 sites surround the
A_2 sites while the B_1 sites are located at the centre of the

rectangular faces of the unit cell.

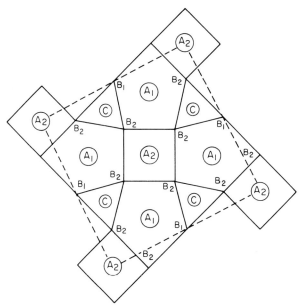

*Fig.2. Schematic representation of the
tungsten bronze structure, in {00l} direction.*

The unit cell may thus be described as $(A_1)_4(A_2)_2C_4$
$(B_1)_2(B_2)_8O_{30}$. A wide variety of substitutions may be made
in the available sites. A_1 and A_2 sites can be occupied by
Na, K, Cs, Ca, Sr, Ba, Pb, rare-earths, etc., and the much
smaller C sites by Li, Be, Mg and Al. B_1 and B_2 sites can
be occupied by Mg, Fe, Ti, Zr, Sn, Nb, Ta, W, etc. Substitutions must be made in such a way that the valence charges of
the ions occupying the A, B and C sites balance those of the
O^{2-} ions and maintain overall electrical neutrality. The
polar axis of the ferroelectric members of this class is
along the <001> axis for all except those containing Pb and
possibly Bi. This implies that only 180° reversal of the
polar direction can take place and only 180° domains can be

formed. Both the non-polar high temperature and polar
phases can be either orthorhombic and/or tetragonal. It is
the single crystal form of the tungsten bronze type ferroelectric compounds which has been primarily studied for both
ferroelectric and electro-optic properties.

The first compound of this crystal class reported to
be ferroelectric was lead metaniobate, $Pb_5Nb_{10}O_{30}$ (Goodman,
1953), with both orthorhombic a and b axes polar, an
exception to the rule for these compounds. Lead metaniobate
has a high Curie temperature, $843^\circ K$, and was developed as a
piezoelectric ceramic for use over a wide temperature range.
It is an unusual material with low permittivity, moderate
piezoelectric activity and very low mechanical Q. This
last point is a serious drawback for many applications but
is of considerable use in ultrasonic flow detection where it
helps to suppress the phenomenon known as ringing.

A number of solid solutions of lead metaniobate have
been studied in the polycrystalline ceramic form but the
only one of any practical significance is $(Pb_{5-x}Ba_x)Nb_{10}O_{30}$.
The optimum piezoelectric properties are found near a phase
boundary occurring at x = 2, which separates two ferroelectric orthorhombic phases. Subbarao (1960) found that
for x <2 the polarisation was in the <110> direction whilst
for 5> x >2 the polarisation was parallel to the <001>
direction indicating marked influence of the highly
polarisable lead ion on the polar axis. Compositions
around the phase boundary possess a relatively low temperature coefficient of the resonance frequency, high mechanical
Q and moderately strong piezoelectric activity which renders
them suitable for resonant piezoelectric devices requiring
frequency stability with temperature.

More complex compounds exhibiting the tungsten bronze structure were reported by Roth and Fang (1960), Ainger et al.(1970) and Isupov (1964). One of the more interesting compounds, barium gadolinium iron niobate, $Ba_4Gd_2Fe_2Nb_8O_{30}$, was reported to be a ferroelectric and ferromagnetic ceramic, but investigations with similar ceramics showed that the weak magnetic properties were due to the presence of a second phase, barium hexaferrite.

C.   LAYER STRUCTURE OXIDES AND COMPLEX COMPOUNDS

A large number of layer structure compounds of general formula $(Bi_2O_2)^{2+}(A_{x-1}B_xO_{3x+1})^{2-}$ have been reported (Smolenskii et al. 1961; Subbarao, 1962), where A = Ca, Sr, Ba, Pb, etc., B = Ti, Nb, Ta and x = 2, 3, 4 or 5. The structure had been previously investigated by Aurivillius (1949) who described them in terms of alternate $(Bi_2O_2)^{2+}$ layers and perovskite layers of oxygen octahedra. Few have been found to be ferroelectric and include $SrBi_2Ta_2O_9$ (T_c = 583°K), $PbBi_2Ta_2O_9$ (T_c = 703°K), $BiBi_3Ti_2TiO_{12}$ or $Bi_4Ti_3O_{12}$ (T_c = 948°K), $Ba_2Bi_4Ti_5O_{18}$ (T_c = 598°K) and $Pb_2Bi_4Ti_5O_{18}$ (T_c = 583°K). Only bismuth titanate $Bi_4Ti_3O_{12}$ has been investigated in detail in the single crystal form and is finding applications in optical stores (Cummins, 1967) because of its unique ferroelectric-optical switching properties. The ceramics of other members have some interest because of their dielectric properties.

More complex compounds and solid solutions are realisable in these layer structure oxides but none have significant practical application.

D.   PEROVSKITE TYPE FERROELECTRICS

The general formula of this family of compounds is ABO_3, represented by the natural mineral $CaTiO_3$, called perovskite and which has a cubic unit cell.   This is the non-polar phase of the perovskite type ferroelectrics from which three polar psuedo-cubic distortions, tetragonal, orthorhombic and rhombohedral are derived.   These are shown in Fig.3 for barium titanate, $BaTiO_3$, together with the temperatures of the three transitions.   It is found that, (a) in the tetragonal phase any one of six equivalent <100> axes is polar,   (b) in the orthorhombic phase any one of twelve equivalent <110> axes is polar, and (c) in the rhombohedral phase any one of equivalent <111> directions is polar.   This results in more complicated domain arrangements which have relevance and importance in polycrystalline ceramics.

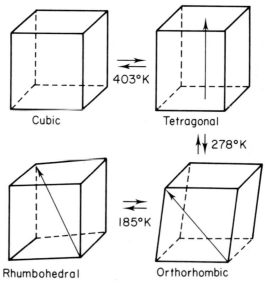

Fig.3.   Perovskite-type distortions for $BaTiO_3$.

Barium titanate has been the most extensively studied member of the perovskite family as a single crystal and has been commercially exploited in the ceramic form.   The single crystals studied were mainly grown from a potassium fluoride flux, which conveniently gave large butterfly twins consisting of triangular plates with <001> axes perpendicular to major faces.   The principal properties of barium titanate are usually based on measurements made on this twinned material.   It is interesting to note that the Curie point is usually quoted as $120^{\circ}C$ whereas that for the corresponding pure ceramics is $130^{\circ}C$ (Roberts, 1949).   More recently, crystals grown from excess titania were shown to have a Curie temperature of $134^{\circ}C$ and distinctly different dielectric properties at very high frequencies.   These differences immediately highlight the role of impurities and source of materials used in crystal growing.   Similar effects will be noted later in the ceramic form.   A review of barium titanate is given by Jona and Shirane (1962).

The discovery of ferroelectricity in barium titanate initiated a search for other perovskite type ferroelectrics which has resulted in improved materials for use as dielectrics and piezoelectrics in a range of devices.

A large number of compounds of the type ABO_3 can be formed.   A can be mono, di and trivalent and B be tri, quadri, penta and hexavalent.   Even more combinations are possible if oxygen is also replaced.   For instance, a large number of double halides belong to the perovskite family but as yet only one, $CsGeCl_3$, has been found to be ferroelectric (Christensen et al. 1965).   In any search for new materials some sort of rationale has to be introduced in the form of a systematic study.   In the present case two questions must be

answered, (a) how likely is a particular compound to
possess the perovskite structure, and (b) what factors
govern the occurrence of ferroelectricity in this lattice.
The answer to (a) can be considered on the basis of ionic
size since the packing of ions in the lattice is important.
Perovskite has a three dimensional network of BO_6 octahedra
with cubic close packed A and O ions and B filling the
interstices.   The packing of this lattice can be characterised by a tolerance factor 't' which is defined by
$R_A + R_X = t\sqrt{2} (R_B + R_X)$ and the radii ratio rules
$R_A/R_X > 0.73$ and $0.41 < R_B/R_X < 0.73$, where R_A, R_B and R_X are the
ionic radii of A, B and X ions respectively.   When $t = 1$,
the packing is ideal but as is seen from Table I (tolerance
factors for ABX_3 perovskite compounds), $t = 0.935$ for $BaTiO_3$.

TABLE I

Compound	R_A	R_B	R_X	t	R_A/R_X	R_B/R_X
$CaTiO_3$	0.99	0.68	1.40	0.8125	0.7071	0.4857
$CaSnO_3$	0.99	0.71	1.40	0.8010	0.7071	0.5071
$SrTiO_3$	1.13	0.68	1.40	0.8602	0.8071	0.4857
$SrZrO_3$	1.13	0.80	1.40	0.8133	0.8071	0.5714
$SrSnO_3$	1.13	0.71	1.40	0.8480	0.8071	0.5071
$BaTiO_3$	1.35	0.68	1.40	0.935	0.9643	0.4857
$BaSnO_3$	1.35	0.71	1.40	0.9217	0.9643	0.5071
$CdTiO_3$	0.97	0.60	1.40	0.8060	0.6929	0.4857
$PbTiO_3$	1.20	0.68	1.40	0.8840	0.8571	0.4857
$PbZrO_3$	1.20	0.80	1.40	0.8355	0.8571	0.5714
$CsGeCl_3$	1.69	0.93	1.61	0.9188	1.0497	0.5776
$CsPbCl_3$	1.69	1.20	1.61	0.8306	1.0497	0.7453

The occurrence of ferroelectricity in $BaTiO_3$ was thought to
be related to the rattling of a Ti^{4+} ion and was supported

TABLE II

Crystal	T_c (°K)	P_s (10^{-6} C/cm^2)	$\varepsilon_\parallel$ Polar Axis	K_p (300°K)	Ferroelectric Phase Structure
$NaNbO_3$	73	12 (373°K)	2000 (73°K)		Monoclinic
$KNbO_3$	708	22 (373°K) 30 (473°K)	530 (373°K) 2000 (475°K)	0.28	Tetragonal Orthorhombic Rhombohedral
$NaTaO_3$			Ferroelectric		
$KTaO_3$	13				
$SrTiO_3$	28		22500 (10°K)		Tetragonal
$BaTiO_3$	393 307	26 (300°K)	150 (300°K)	0.47	Tetragonal Orthorhombic Rhombohedral
$PbTiO_3$	760		120 (300°K)		Tetragonal
$Pb(Fe_{1/2}Nb_{1/2})O_3$	387		2000 (300°K)		Rhombohedral
$Pb(Mg_{1/3}Nb_{2/3})O_3$	265	24 (100°K)	4000 (210°K)	0.16	Distorted Perovskite
$Pb(Fe_{1/2}Ta_{1/2})O_3$	233	28 (120°K)	3000 (300°K)		Rhombohedral

TABLE III

Ceramic	Dielectric Constant	T_c °K	Resistivity Ωm	Coupling Factor K_p	Piezo-Constant d_{31} C/N	P_s 10^{-6} C/cm^2	Comments
$Na_{\frac{1}{2}}K_{\frac{1}{2}}NbO_3$	400-290	693	10^{10}	0.40	-32×10^{-12}	30	Hot pressed or air fired and are mainly used for high frequency transducers
$BaTiO_3$ (Casonic)[+]	1100	383	10^9	0.26	-50×10^{-12}	6	Modified with PbO, MgO and CoO to give properties for power transducers
$BaTiO_3$	1500	385	10^{10}	0.37	-75×10^{-12}	8	
PZT*2	450	643		0.47	-60 "	40	
" 4	1300	600		0.58	-123 "	30	
" 5A	1700	638		0.60	-171 "	38	
" 5H	3400	466		0.65	-274 "	33	
" 6A	1100	608		0.39	-77 "	30	
" 6B	500	623		0.25	-29 "	15	
" 7A	425	623		0.52	-61 "	42	
" 8	1000	573		0.50	-95 "	25	
$Pb_5Nb_{10}O_{30}$	225	843		0.07 $(k_{33}=0.33)$	$d_{33}=85$ "	-	Not a perovskite but tungsten bronze structure type

[+]Plessey Trade Mark.   *Clevite Trade Mark.

by the fact that substituting either Ca^{2+} and Sr^{2+} for Ba^{2+}
or Sn^{4+} and Zr^{4+} for Ti^{4+} lowered the Curie temperature with
increasing amounts of any one substitution and thereby
reduced the rattling of Ti^{4+}. However there are exceptions
to the rule and the substitution of Pb^{2+} for Ba^{2+} increases
T_c. Other factors such as polarisability and bond
character have therefore to be considered.

Further subdivisions of the A and B ions have been
considered, mainly by Smolenskii et al. (1958, 1959), giving
rise to complex perovskite type compounds of the general
type $(A'_{1-x}A''_x)(B'_yB''_{1-y})O_3$. Hence a large number of solid
solutions can be made in the perovskite family of compounds
from which materials may be selected for specific applications. Most of the effects observed in such materials have
been derived empirically on polycrystalline ceramic forms,
single crystals being generally difficult to produce.

A number of the more important perovskite members are
given in Table II (properties of selected ABO_3 single
crystals) and Table III (properties of ferroelectric ceramics
at $298^{\circ}K$) of which lead titanate-zirconate compounds assume
special importance because of their piezoelectric and
electro-optic properties. It is important to note that the
properties are dependent upon the material used for the
actual measurements. Single crystals produced by different
manufacturers using the same technique and purity of raw
materials generally exhibit the same properties. However,
for ceramic specimens the differences in manufacturing technique peculiar to a particular production environment will
be reflected in the properties. Therefore such Tables are
used purely as guides to the kind of properties expected and
differences may arise from source to source. The properties

of the ceramics indicate how a basic composition can be
modified through solid solution effects to optimise a
particular property.

III.   FERROELECTRIC CERAMICS

The materials to be considered have a typical ceramic
structure consisting of small crystallites.   All ferroelectrics have within their single crystals or crystallites
a further sub-division into domains, each domain being a
small region of uniform saturation polarisation with the
direction of polarisation confined to a particular crystallographic axis.   Although polarisation within a domain has
the maximum saturation value, in the virgin state the
crystallites or ceramic elements have no overall polarisation.
This arises because the direction of polarisation within a
crystallite and the directions of the crystallite axes are
random.   The direction of polarisation within the domains
can be changed by applying a strong electric field to the
material.   This reverses the direction of polarisation
within a particular domain in some materials and rotates it
in others.   The electric field necessary to produce these
two types of polarisation are generally different.   The
reversal (or 180°) processes do not involve any change in
dimensions in contrast to those involving rotation.   These
two types of domain switching in ferroelectric crystals have
different associated coercive fields.   Re-orientation by
180° does not involve strain, whilst re-orientation by 90°
(tetragonal, orthorhombic), 60-120° (orthorhombic) or 71-109°
(trigonal) is accompanied by strains and resultant internal
stresses.   During polarisation by the application of a
large DC field, substantial strains and changes in dielectric

constant occur and a remanent polarisation is induced.
Subsequently when the ceramic is subjected to mechanical
or electrical AC stresses, charge is released and changes in
both strain and dielectric constant occur.   The polarised
ceramics respond to axial compression in a similar way for
both static and transient ($\sim$ milliseconds) loading conditions.
The open circuit field generated is proportional to the
axial stress, whilst under short circuit conditions domain
re-orientation occurs even at fairly low stress levels, so
that the response is both ferroelectric and piezoelectric.
Selected ceramics can be utilised repeatedly for releasing
charges at defined voltages provided that the stress level
is not excessive, e.g. lead-zirconate-titanate mixtures.

A.   PREPARATION

Ceramics are usually made by solid state reaction of
the constituent oxides or carbonates.   The constituents are
intimately mixed and fired or calcined at a temperature some
$300^{\circ}C$ below their final sintering temperature.   The extent
of the reaction is checked by X-rays, after which the
material is finely ground and mixed with an organic binder
to facilitate the pressing of the desired shape.   The
binder is burnt off and the ceramic pieces finally fired
under empirically determined conditions of time and temperature in order to produce optimum properties.   The density,
porosity and X-ray structure are subsequently determined,
after which the material is cut and lapped to the final
dimensions.   Electrodes, usually made of silver, are fired
on and after impregnation with a silicone oil the ceramics
are poled with a DC field, again under empirically optimised
conditions of temperature and time.   Most piezoelectric

TABLE IV

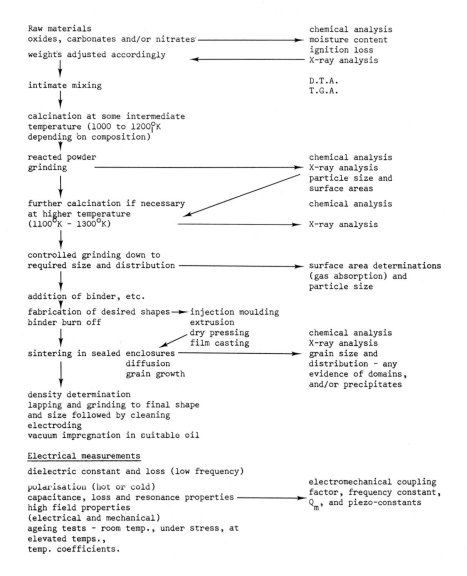

Raw materials
oxides, carbonates and/or nitrates ——————————→ chemical analysis
                                                moisture content
weights adjusted accordingly   ←——————————— ignition loss
                                                X-ray analysis
  ↓
intimate mixing   D.T.A.
  ↓   T.G.A.

calcination at some intermediate
temperature (1000 to 1200°K
depending on composition)
  ↓
reacted powder   chemical analysis
grinding ——————————————————————————→ X-ray analysis
  ↓   particle size and
                                                surface areas
further calcination if necessary   chemical analysis
at higher temperature
(1100°K - 1300°K)   ——————————————————————→ X-ray analysis
  ↓

controlled grinding down to
required size and distribution ————————————————→ surface area determinations
  ↓   (gas absorption) and
                                                particle size
addition of binder, etc.
  ↓
fabrication of desired shapes —→ injection moulding
binder burn off   extrusion
  ↓   dry pressing   chemical analysis
                                film casting   X-ray analysis
sintering in sealed enclosures ———————————————→ grain size and
  ↓   diffusion   distribution - any
                grain growth   evidence of domains,
  ↓   and/or precipitates
density determination
lapping and grinding to final shape
and size followed by cleaning
electroding
vacuum impregnation in suitable oil

Electrical measurements

dielectric constant and loss (low frequency)

polarisation (hot or cold)   electromechanical coupling
capacitance, loss and resonance properties ————→ factor, frequency constant,
high field properties Q_m, and piezo-constants
(electrical and mechanical)
ageing tests - room temp., under stress, at
elevated temps.,
temp. coefficients.

ceramics are solid solutions where variation of chemical
composition allows optimisation of desired properties.

B.   EFFECTS OF CERAMIC PROCESSING

Ceramic preparation or processing can have profound
effects upon the microstructure and electrical properties of
a given material.   Furthermore, for the lead titanate
zirconate solid solutions, the atmosphere used for sintering
also exerts an influence.   A schedule for the processing of
lead titanate-zirconate ceramics is given in Table IV from
which several important factors emerge.   These are (1) the
purity, crystal form and particle size of the raw materials,
(2) the effectiveness of the mixing process and possible
contamination, (3) the reactivity of mixed materials, the
lowest temperature for the formation of solid solution and
the minimisation of the loss of volatile components, e.g.
PbO, (4) the grinding method and control of impurity pick
up and particle size, (5) the final firing conditions
(temperature, time and atmosphere) and the formation of
homogeneous single phase material with a required grain size
and distribution.   The dielectric and electromechanical
properties are further affected by the poling technique and
this has to be empirically derived for a particular
composition.

Since actual process details are generally confidential,
it is difficult to know which stages have the most profound
effect.   The following examples are from experiments
carried out with barium titanate and lead titanate-zirconate
and show some influences of a specific impurity, iron oxide.

1. Barium titanate

In the development of commercial ceramics for both
dielectric and piezoelectric devices it was found that the
cubic-hexagonal transition (1460°C) was inhibited.   It was
noted that iron oxide addition at the 1 mol.% level has two
distinctive effects depending upon the temperature used for
sintering the ceramic.   Firstly, at lower temperatures a
non-polarisable ceramic is produced with a very suppressed
Curie peak at 390°K, and secondly, a polarisable, hence
piezoelectric body can be formed with a lower Curie temperature, $T_c \sim 350^{\circ}$K.   The latter material develops the normal
ferroelectric-tetragonal phase with a somewhat lower axial
ratio (c/a) whilst the former occurs as a variable low axial
ratio tetragonal (v.l.a.r.t.) form of $BaTiO_3$.   In the study
of pure barium titanate ceramics prepared from a precipitated
barium titanyl oxalate, Ainger and Herbert (1959) found that
the cubic-hexagonal transition occurred at 1460°C but was
lowered by additions of transition metal oxides, e.g. Fe_2O_3.
However the v.l.a.r.t. modification predominated up to this
transition temperature and no piezoelectric activity could
be induced.   It was also observed that minor additions of
iron oxide improved the quality and size of the butterfly
twins obtained during crystal growth from molten KF.   At
higher concentrations (1 to 2 mol.% Fe_2O_3) solid solution
effects were observed in that the Curie temperature was
reduced by about 60°K/mol.% Fe_2O_3 added.

It would appear that structures which inhibit the full
development of tetragonal crystals are present in barium
titanate formed at low temperatures and they can be
stabilised at higher temperatures by the presence of iron
oxide.   The factors influencing the stability of these

inhibiting structures are complex and include the mode of
formation and purity of the titanate.   Furthermore, the
lowered cubic-hexagonal transition temperature for the high
purity titanate containing iron oxide prohibits the development of the normal tetragonal phase.   However, this
investigation produced a high dielectric constant ceramic
with a low temperature coefficient suitable for stable
capacitor bodies and possessing very low ageing behaviour
since the usual domain re-alignment which occurs with time
was much reduced.

2. Lead titanate zirconate

       During the development of a lead titanate zirconate
mixture, similar to the Clevite PZT4, it was found that the
type of milling employed greatly influenced the resistance
to depolarisation of the poled ceramic.   This particular
PZT was required for power transducers which are subjected
to both electrical and mechanical stresses.   It therefore
required high electrical and mechanical Qs to survive such
conditions without depoling.   Again it was found that
impurity pick-up, particularly during the final milling
stage, was affected by the type of mill and grinding media
used.   Dense grinding media were required since the density
of the PZT slurry was relatively high and steel was an
obvious choice.   Earlier experiments indicated that fairly
long milling times were necessary to produce a sufficiently
fine grain powder which readily densified to a strong
ceramic, exhibiting a tight grain structure.   The piezoelectric as well as loss properties were found to be a
function of the final milling and the iron impurity pick-up.
These results influenced the choice of the final milling

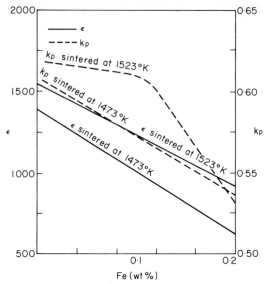

Fig. 4.   *Variation of* ε *and* K_p *with Fe content*
for $Pb_{0.94}Sr_{0.06}Ti_{0.47}Zr_{0.53}O_3$.

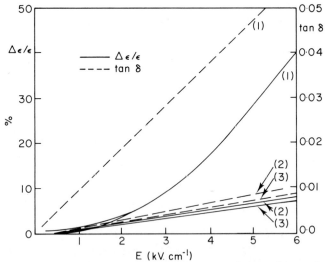

Fig. 5.   *% change in* ε *and* tan δ *with E (1 KHz)*
for $Pb_{0.94}(SrTiZr)_{0.6}O_3$ *(as Fig. 4) containing*
1) 0.08% Fe,   2) 0.20% Fe,   3) 0.33% Fe.

process and a fluid energy (Jet) mill was adopted to avoid contamination of the powder. Further experiments with conventional ball mills, a vibro-energy mill and the Jet mill highlighted the influence of iron oxide and probably grain size (surface area) on the dielectric and piezo-electric properties, as seen in Figs.4 and 5. Obviously the PZTs containing iron oxide have distinctive properties so it would be a dangerous practice to allow a milling process to introduce it by attrition. Hence a Jet mill is used and the required amount of iron oxide added to the PZT.

IV. APPLICATIONS

A. SELECTION OF CERAMIC FOR TRANSDUCER APPLICATIONS

A large number of solid solution effects can be readily made in polycrystalline ceramics which allow prop-erties to be tailored to a given application.

Lead titanate-zirconates have virtually taken over the role of barium titanate in electromechanical transducers. In this system, a complete solid solution series exists from ferroelectric $PbTiO_3$ to antiferroelectric $PbZrO_3$. At about 45-47 mol.% $PbTiO_3$ a morphotropic phase boundary separates the tetragonal and rhombohedral forms. Hereabouts, both dielectric and electromechanical properties reach their maximum values and hence many of the compositions now commercially available are centred on this region. Using the Clevite trademark, the family of compounds have the following properties: PZT2 is a straightforward PZT with a high coupling factor, high Curie temperature and is suitable for shear mode resonance applications. PZT4 is modified with Ca, Sr or Ba to give increased dielectric constant whilst maintaining other excellent properties and is used

for high power resonant transducers and high voltage
generators for spark ignition.   PZT5 is either PZT2 or PZT4
with additions of Bi_2O_3, La_2O_3, Ta_2O_5 or Nb_2O_5.   It
possesses high charge sensitivity, high coupling factor and
lower mechanical Q.   It is used for non-resonant sensing
applications such as gramophone pick-up elements, hydro-
phones, microphones and low power transducers.   PZT6 is
modified with Cr_2O_3 and contains more $PbTiO_3$ to give low
ageing time stability and low temperature coefficient of
resonant frequency.   This high stability PZT has many
applications for r.f. filtering (typically at 450 KHz) and
more recently in energy trapping mode devices for operation
at 10.7 MHz.   PZT8 is modified PZT4 with transition metal
oxide additions and has very low elastic and dielectric
losses with some degradation of dielectric constant and
coupling factor.   It extends the power range for high power
sonic or ultrasonic transducers used in sonar and ultrasonic
cleaning and can be used at high electrical and mechanical
stresses without depoling effects taking place.

A comprehensive review of the applications of piezoelectric materials is given by Jaffe and Berlincourt (1965).

B.   ELECTRO-OPTIC CERAMICS

In the development of a special high charge generator
material for projectile detonation systems a composition
$Pb(Ti_{0.35}Zr_{0.65})O_3$ containing one to two weight percent
Bi_2O_3 or La_2O_3 possessed suitable properties.   In order to
obtain optimum properties, Haertling (1964) used a hot
pressing technique to give ceramics with densities of
greater than 99% of the theoretical value.   It was subsequently found that such materials showing a typical ceramic

structure were transparent in thin sections.   It was shown
by Land et al. (1969) that the effective birefringence was
dependent upon domain alignment which can be incrementally
or partially switched by an electric field.   Further
materials research has improved transparency.   The 100%
transmission at a thickness of 0.012cm in the lanthanum
substituted PZTs (described by Haertling and Land, 1971)
provides exciting possibilities in the field of electro-optic
display and storage devices.   Full descriptions of these
materials and devices are reported by Land and Holland (1970)
and by Meitzler et al. (1970).

C.   PYROELECTRIC CERAMICS

Polarised ferroelectric ceramics exhibit remanent
polarisations, the values of which depend upon the
composition, Curie temperature (T_c), thoroughness of poling
and sample preparation.   Barium titanate ceramics have
values of $P_r \sim 5 - 8 \times 10^{-6} C.cm^{-2}$ whilst those of the lead
titanate zirconates are much higher, $30 - 40 \times 10^{-6} C.cm^{-2}$.
Operating a ferroelectric near to the Curie temperature
results in high values of the pyroelectric coefficient
which are temperature dependent.   Likewise, the dielectric
constant changes more rapidly near T_c.   In applications of
pyroelectricity for infra-red detection a high pyroelectric
coefficient and low dielectric constant are required.
Therefore, a compromise has to be made and solid solution
effects may be utilised to change T_c.   In barium titanate
T_c may be moved near to room temperature by the substitution
of strontium for barium.   For the lead titanate-zirconate
system more extensive solid solution effects may be used
and advantage taken of the high values of remanent

polarisation available.

The most common means of lowering T_c in this system is by the substitution of La, Ba or Sr for Pb, and Sn for Ti and Zr. The lead titanate zirconates exhibit the largest figures of merit for ferroelectric ceramics and offer flexibility in detectors based on a pyroelectric ceramic.

V.   SUMMARY

It has been shown that the development of ferroelectric ceramics has created new opportunities for the device engineers as well as a basis for more thorough research into the behaviour of ferroelectrics in the single and poly-crystalline forms. The common feature of these oxide bodies is the BO_6 octahedra. The arrangement of the octahedra in a three dimensional crystal lattice, together with the filling of the interstitial sites thus formed, have profound effects on the ferroelectric properties. The major break-through in these ferroelectric oxides came with the discovery of the lead titanate-zirconate system which has been shown to exhibit a wide range of properties suitable for many transducer types and electro-optic applications. However, the processing procedures are every bit as important as the chemical compositions themselves and careful control is necessary to reproduce the required properties.

VI.   APPENDIX

Parameters used in the text and figures are defined as follows:

1. Coupling factor k, where $k^2 =$

   $\dfrac{\text{stored electrical or mechanical energy}}{\text{total stored energy}}$, but does not

   necessarily represent the overall efficiency.

2. ε = dielectric constant.

3. Tan δ = dielectric loss.

4. Piezoelectric strain constant, d, is

   $\dfrac{\text{charged developed}}{\text{applied mechanical stress}}$ $\left(\dfrac{C/m^2}{N/m^2}\right)$

5. Piezoelectric strain constant, g, is

   $\dfrac{\text{field developed}}{\text{applied mechanical stress}}$ $\left(\dfrac{V/m}{N/m^2}\right)$

6. Mechanical Q, Q_m, gives some measure of the internal friction and the higher Q_m the lower is the internal friction.

REFERENCES

Ainger, F.W., Bickley, W.P. and Smith, G.V. (1970) Proc. Brit. Ceram. Soc. 221.

Ainger, F.W. and Herbert, J.M. (1959) Trans. Brit. Cer. Soc. 58, 410.

Aurivillius, B. (1949) Arkiv Kemi 1, 463 and (1950) 2, 519.

Christensen, A.N. and Rasmussen, S.E. (1965) Acta. Chemica Scandinavica 19, 421.

Cummins, S.E. (1967) Proc. I.E.E.E. 55, 1536 and 1537.

De Bretteville Jr., A. (1946) Bull. Amer. Phys. Soc. 21, 18.

DiDomenico Jr., M. and Wemple, S.H. (1968) Appl. Phys. Lett. 12, 352.

DiDomenico Jr., M. and Wemple, S.H. (1969) J. Appl. Phys. 40, 720.

Goodman, G. (1953) J. Amer. Ceram. Soc. <u>36</u>, 368.

Haertling, G.H. (1964) Amer. Ceram. Soc. Bull. <u>43</u>, 875.

Haertling, G.H. and Land, C.E. (1971) J. Amer. Ceram. Soc.
<u>54</u>, 1.

Hulm, J.K. (1953) Phys. Rev. <u>92</u>, 504.

Isupov, V.A. (1964) Sov. Phys. Crystallography <u>9</u>, 281.

Jaffe, H. and Berlincourt, D.A. (1965) Proc. I.E.E.E. <u>53</u>,
1372.

Jamieson, P.B., Abrahams, S.C. and Bernstein, J.L. (1968)
J. Chem. Phys. <u>38</u>, 5066.

Jona, F. and Shirane, G. (1962) Ferroelectric Crystals,
Chapter IV, Pergamon Press.

Land, C.E. and Holland, R. (1970) I.E.E.E. Spectrum <u>7</u>, 71.

Land, C.E. and Thacher, P.D. (1969) Proc. I.E.E.E. <u>57</u>, 751.

Meitzler, A.H., Maldonado, J.R. and Fraser, D.B. (1970)
B.S.T.J. <u>49</u>, 953.

Roberts, S. (1949) Phys. Rev. <u>75</u>, 989.

Roth, R.S. and Fang, P.H. (1960) Bull. Amer. Phys. Soc.
(Ser.2) <u>5</u>, 58.

Sawyer, C.B. and Tower, C.H. (1930) Phys. Rev. <u>35</u>, 269.

Smolenskii, G.A. and Agranovskaya, A.I. (1958) Zhur. Tekh.
Fiz. <u>28</u>, 1491.

Smolenskii, G.A., Agranovskaya, A.I. and Popov, S.N. (1959)
Fiz. Tverdgo-Tela, <u>1</u>, 167.

Smolenskii, G.A., Agranovskaya, A.I., Popov, S.N. and
Isupov, V.A. (1958) Zhur. Tekh. Fiz. <u>28</u>, 2152.

Smolenskii, G.A., Isupov, V.A. and Agranovskaya, A.I. (1959)
Fiz. Tverdgo-Tela <u>1</u>, 170.

Smolenskii, G.A., Isupov, V.A. and Agranovskaya, A.I. (1961)
Sov. Phys. Sol. State <u>3</u>, 651.

Subbarao, E.C. (1960) J. Amer. Ceram. Soc. <u>43</u>, 439.

Subbarao, E.C. (1962) J. Phys. Chem. Solids 23, 655.

Valasek, J. (1921) Phys. Rev. 17, 475.

Wul, B.M. and Vereshchagin, L.F. (1945) Compt. Rend. Acad. Sci. U.R.S.S. 48, 634.

MAGNETIC OXIDES

D.J. MARSHALL

Royal Radar Establishment,
Malvern, Worcesterhsire

I.   INTRODUCTION

II. CRYSTAL STRUCTURES
        A. SPINEL FERRITES
        B. MAGNETO-PLUMBITES
        C. GARNETS
        D. ORTHOFERRITES

III. APPLICATIONS
        A. PERMANENT MAGNETS
        B. COMMUNICATIONS COMPONENTS
        C. MICROWAVE COMPONENTS
        D. COMPUTER MEMORIES
        E. MAGNETIC RECORDING SYSTEMS
        F. POTENTIAL APPLICATIONS

IV. PREPARATION
        A. POLYCRYSTALLINE CERAMICS
        B. SINGLE CRYSTALS
        C. GROWTH OF THIN LAYERS

I.   INTRODUCTION

Magnetic oxides provide the basis for many vital components. Their uses range from latches for cupboard

doors, through components which reduce the size and cost of
television receivers and permit increased sophistication in
radar and communication systems, to the magnetic toroid
which provides the memory store for most modern computers.
Magnetic oxides are increasingly used as permanent magnets
to produce cheap and compact electric motors and there is a
prospective use in cheap, robust, mini-computers.

The basic property of magnetic oxides which makes them
useful was named ferrimagnetism by Néel (1948).   This
effect arises because the angular momentum of an electron
resulting from its spin or orbital motion gives rise to a
magnetic moment.   In most oxides the electron shells are
full and all the electrons are paired so these effects
cancel out, resulting in a paramagnetic material.   In the
case of oxides involving transition metals (Mn, Fe, Co, Ni)
or the rare-earths, partially filled electron shells occur
below the valence shells and the materials contain unpaired
electrons with associated magnetic moments.   As the temperature of such solids is lowered there comes a point at
which the magnetic moments become aligned in two or more
sub-lattices.   This is due to super-exchange forces arising
from the electron interactions induced by the overlap of the
orbitals of the oxygen and metal atoms.   In simple cases,
e.g. MnO and NiO, two oppositely orientated sub-lattices,
each containing the same number of the same magnetic ions,
result in an antiferrimagnetic structure.   If the sub-
lattices contain different numbers of magnetic ions, or ions
with different magnetic moments, or if the sub-lattices are
orientated at an angle to each other, then the material will
have a net macroscopic magnetic moment (Rado and Suhl, 1963;
Brailsford, 1966).   The four most important structures into

which magnetic oxides can be grouped are considered here.

II.   CRYSTAL STRUCTURES

A.   SPINEL FERRITES

This group has the general formula MFe_2O_4 where M is a divalent ion such as Mg, Mn, Fe, Co, Ni, Cu, Zn or Cd.   Li can also be incorporated into spinel ferrites provided that the charge is balanced by the simultaneous incorporation of a ferric ion ($M = Li^+_{0.5}Fe^{3+}_{0.5}$).   Trivalent iron atoms in the spinel structure may also be replaced by other trivalent metals such as Al^{3+}, Ga^{3+} or Cr^{3+}.   Since each of these ferrites crystallise in the same cubic structure, extensive substitutions and a wide range of solid solutions are possible.

In the unit cell of a spinel ferrite there are 32 oxygen atoms packed to leave two types of site for the metal ions.   There are 64 sites which are tetrahedrally surrounded by four oxygens.   These are called A sites, 8 of which are occupied by metal ions.   The other 32 sites, called B sites, are octahedrally surrounded by six oxygens, and 16 of these sites are occupied.   Some of the divalent ions which can form spinel ferrites have a strong preference for the A site (Mn, Zn), some for the B site (Co, Ni), whilst others have no marked preference.   Ferrites can thus be classified into two types of structure;   the normal spinel in which 8 divalent ions enter the A sites and 16 trivalent ions enter the B sites, and the inverse spinel in which the B sites are shared between 8 divalent ions and 8 trivalent ions, with the remaining 8 trivalent ions going onto the A sites. These two situations are the ideal end members of a series within which various degrees of inversion and order are shown, depending upon the composition and conditions of preparation (Smit and Wijn, 1959).

The saturation magnetisation of a ferrite is strongly affected by the magnetic moment of the ions present and the sites on which they are located. This is illustrated by two simple examples.

Zinc has a strong preference for the tetrahedral site and so $ZnFe_2O_4$ is a normal spinel, with all magnetic atoms in the B sites. Usually the super-exchange interaction between the A and B sub-lattices is the dominating factor in the alignment of the atomic magnetic moments, but since zinc has no unpaired electrons there is no AB interaction and the much weaker BB interaction forces the spins of the ferric ions into an antiparallel alignment with no resultant magnetic moment. On the other hand, nickel prefers an octahedral co-ordination so that $NiFe_2O_4$ forms an inverse spinel. In this case the eight tetrahedral sites are occupied by ferric ions, each having five unpaired electrons whilst the octahedral sites contain eight Fe^{3+} ions with five unpaired electrons each and eight Ni^{2+} ions with two unpaired electrons each. The stronger AB interaction takes effect causing the spins on the A sites to align themselves antiparallel to the spins on the B sites, resulting in a net magnetic moment.

If a solid solution of nickel-zinc ferrite is formed, zinc atoms displace iron atoms from the A sites forcing them onto the B sites and increasing the imbalance between the magnetic moments of the two oppositely aligned sub-lattices. As nickel is replaced by zinc the saturation magnetisation of the material increases. When more than half the nickel atoms have been replaced by zinc the AB interaction is

weakened and cannot maintain the anti-parallel alignment of
the two sub-lattices, hence the BB interaction begins to
take over and the saturation magnetisation decreases.   The
substitution of zinc into the tetrahedral sites also lowers
the Curie temperature since the temperature dependence of
the A sub-lattice magnetisation is less pronounced than that
of the B sub-lattice.   The dilution of the A sub-lattice
magnetisation and reduction of its super-exchange interaction strongly increases the variation of the magnetic
properties with temperature.

B.   MAGNETO-PLUMBITES

      The crystal structures of magneto-plumbites are
hexagonal, and magnetic compounds in this class are often
known as hexaferrites, e.g. barium ferrite, $BaFe_{12}O_{19}$.
Barium can be replaced by strontium, lead, or some of the
larger rare-earth ions.   Other related structures occur in
this group, e.g. $Ba_2Me_2Fe_{12}O_{22}$, where M is a divalent metal.
All these structures are composed of three basic building
blocks, termed the S, R and T blocks.   In the S block, the
oxygen anions and the interstitial cations are arranged as
in the spinel structure viewed with the {111} axis vertical.
The unit formula is Fe_6O_8 with four cations in octahedral
sites and two in tetrahedral sites.   The R block, $BaFe_6O_{11}$,
consists of three layers of oxygen atoms but with an oxygen
in the middle layer replaced by barium.   Five cations are
in octahedral co-ordination, but distributed between two
different types of site, whilst the sixth cation lies in
the same plane as the barium atom in a third type of site
which has five-fold oxygen co-ordination.   The magneto-
plumbite (or M) structure is built up from these two blocks

stacked in the sequence ..RSR*S*.. (the asterisk indicates
that the unit has been rotated through 180° in the basal
plane).   The T block, $Ba_2Fe_8O_{14}$, has four layers of oxygen
atoms.   One oxygen position in each of the middle two
layers is filled by a barium atom and the cations are distributed between six octahedral and two tetrahedral sites,
with their spins so aligned that if all have the same
magnetic moment the resultant moment of the block is zero.
The Y-type hexagonal ferrites have the structure STSTST, and
contain two divalent cations per formula unit to maintain
electrostatic neutrality, e.g. $Ba_2Zn_2Fe_{12}O_{22}$.   Diagrams of
these structures and further details of possible compositions
are given by Smit and Wijn (1959).   The principal features
of hexagonal ferrites are high saturation magnetisation and
very high uniaxial anisotropy, which result in high coercivity and remanence.   These materials provide very good
permanent magnets at relatively low cost.

C.   GARNETS

     Synthetic magnetic garnets with general formula
$R_3Fe_5O_{12}$ (R = yttrium or rare-earth), crystallise in a cubic
structure containing 96 oxygen atoms in a unit cell.   The
anion packing gives 16 tetrahedral oxygen sites ('a' sites,
similar to A in spinel) 24 larger octahedral sites ('d'
sites, similar to the B site in spinel), and 24 larger still
dodecahedral (or 'c') sites which are surrounded by eight
oxygens.   In yttrium iron garnet (YIG), $Y_3Fe_5O_{12}$, the 40
Fe^{3+} ions fill the a and d sites and the 24 Y^{3+} ions fill
the c sites.   In the pure material each site is occupied in
a completely ordered fashion by an ion of fixed valency and
there is not the same scope for variation as in the spinel

structure.   The structure is also more magnetically dilute
and the super-exchange interactions are weaker, resulting in
much lower Curie temperatures for the garnets.

As for the spinels there is a strong interaction
between the iron atoms on the a and d sub-lattices such that
the magnetic moments on each sub-lattice are aligned parallel
with each other, and antiparallel to those on the other sub-
lattice, resulting in a net magnetic moment equivalent to
one ferric ion per formula unit, if the ion on the c site is
non-magnetic.   The rare-earth ions, except for Y^{3+}, La^{3+}
and Lu^{3+}, do have a magnetic moment due to unpaired f
electrons, and in this case the contribution from orbital
motion is no longer negligible.   The interaction of the c
sub-lattice with the a and d sub-lattices is relatively weak,
but still aligns the net rare-earth moment anti-parallel to
the net ferric moment (that is antiparallel to the d sub-
lattice), at least for the rare-earths from Gd to Yb.   The
temperature dependence of the alignments on the two types of
lattice is quite different, and there is a temperature (the
compensation point) at which the magnetic moments can exactly
cancel out.

The ionic size requirements of the garnet structure
are more rigid than for the spinels and some of the larger
rare-earths (from La to Nd) cannot form a pure garnet,
although they can be accommodated by association with a
smaller rare-earth ion in a mixed garnet.   Substituted
garnets can also be formed in which some of the iron is
replaced, e.g. by Al^{3+} or Ga^{3+} on the 'a' sites, or by In^{3+}
or Sc^{3+} on the 'd' sites.   Some effects which these variations have on the magnetic properties of these garnets will
be mentioned later.

D.   ORTHOFERRITES

Rare-earth orthoferrites have the general formula
$RFeO_3$, where R can be yttrium or one of the rare-earths.
They crystallise in a perovskite-like orthorhombic structure
with the ferric ions occupying octahedral sites surrounded
by six oxygens, and the larger rare-earth ions occupying a
distorted site surrounded by about 9 nearest-neighbour
oxygens rather than the symmetrical 12 co-ordinated site in
the ideal perovskite structure (Geller, 1956).   The spins
of the ferric ions are ferrimagnetically aligned on two
oppositely orientated sub-lattices, but are canted such that
there is a small net magnetic moment of $\sim$ 100gauss.   This
moment lies along the c-axis in all the orthoferrites except
$SmFeO_3$, in which it lies along the a-axis.   The weak interaction between the iron sub-lattices and the magnetic
rare-earth ions gives complicated magnetic properties (Treves
1961 and White, 1969).   More recently this group of
compounds has aroused interest because their high anisotropy
and low magnetisation saturation enable them to support
large bubble domains.

III.   APPLICATIONS

A.   PERMANENT MAGNETS

A useful permanent magnet material must have a high
saturation magnetisation and a very high coercive force, and
such materials are characterised by a figure of merit, the
energy product, $(BH)_{max}$.   Metallic magnets with energy
products up to 10^7 gauss-oersted are available, but the
cheaper barium ferrites can be made with $(BH)_{max} \sim 10^6$ in
isotropic ceramics, which can be further increased to $\sim$
4×10^6 in anisotropic ceramics with grains having a high

degree of orientation.   Ceramic magnetic oxides now provide
about a quarter of all permanent magnets used and the
proportion is still increasing.   Besides such mechanical
uses as latches, chucks or separators the most important
applications are for small DC motors and light compact
sources of biasing magnetic fields (Bozorth, 1969; Jacobs,
1969).

B.   COMMUNICATIONS COMPONENTS

This category covers the use of soft ferrites in radio,
television and telephone systems.   A soft magnetic material
has a small area hysteresis loop and high ·initial permeability, e.g. manganese-zinc and nickel-zinc ferrites.
These uses account for the greatest bulk of ferrite produced
since in every television set there is at least a ferrite
based deflection transformer and a deflection yoke.   These
ferrites are also very widely used as cores for inductors
and transformers where their very low eddy current losses
make them preferable to metallic cores.

C.   MICROWAVE COMPONENTS

Modern radar and communication systems operating at
frequencies >1 GHz use components such as isolators, circulators, switches and filters which rely on ferrites for their
efficient operation (Harvey, 1963).   These devices are
based on aspects of the gyromagnetic effect in magnetic
oxides which is related to the precessional motion of the
spinning electron in a magnetic field.   For example, the
plane of polarisation of high frequency electromagnetic
radiation is rotated by over $100°/cm$ (the Faraday effect) as
it passes through certain ferrites, such as $(Mg, Mn)Fe_2O_4$.

Since this effect is non-reciprocal it can be made the basis
of isolators and circulators both in conventional waveguides
and in more recent stripline configurations.   Another
result of the precession of the electron is the resonance
which occurs when the frequency of the microwave field
matches the precession frequency in the material.   This
frequency is governed by the applied magnetic field as modified by various internal demagnetising fields so that if the
power absorbed at some given microwave frequency is plotted
against the applied field then a peak of the type shown in
Fig.1 will be obtained.   The width of the peak at half
height (or 3dB down from max.) is known as the resonance
linewidth (ΔH) of the material and may vary from several
hundred oersteds in a polycrystalline spinel ferrite to a
fraction of an oersted in single crystal YIG.   It is a
measure of the rate at which energy is dissipated into the
surrounding crystal.   This effect is also non-reciprocal in
that only microwave radiation whose direction of circular
polarisation is in the same sense as the direction of precession will be coupled into the crystal, and so it can be used
as the basis for circulators, isolators and filters as well
as power limiters (Lax and Button, 1962;   Hudson, 1970).

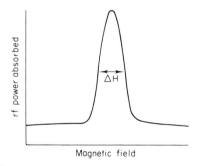

Fig.1. Ferrimagnetic resonance in a microwave ferrite.

D.   COMPUTER MEMORIES

The majority of computer main stores are based on square loop ferrite toroids.   The basic material is usually Mn-Mg spinel ferrite doped with proprietary additions. Since their inception the size and cost of cores has decreased and the speed of operation has increased; a current read-write cycle time is about 200 ns (Richard, 1970).

This application is based on the ability to produce these particular spinel ferrites with a very rectangular hysteresis loop (Fig.2).   The application of a magnetic field slightly greater than the coercive force, produced by sending a current along a wire passing through the toroid, will switch the magnetisation in the core from its remanent value in one direction to its remanent value in the opposite direction, thus providing a bistable device, i.e. with 0 and 1 states.

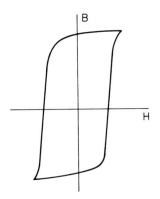

Fig.2.   B-H loop for a 'square loop' ferrite.

E.   MAGNETIC RECORDING SYSTEMS

Probably the largest single application of magnetic oxides, is their use for the recording and storage of information.   Magnetic recording tape is based on the

incorporation of particles of a magnetic oxide into a suitable plastic tape. Magnetic properties required for this
application are a moderate value of the coercivity (a few
hundred oersteds) so that demagnetisation problems are
minimised, a remanent induction of about 1000gauss to ensure
a useful signal amplitude, and a square hysteresis loop for
good resolution of the recording medium. With control of
purity and particle size and shape, γ-Fe_2O_3 meets these
requirements and is the most commonly used medium. For
tapes, acicular particles of CrO_2 have advantages in some
applications and for discs metallic films of cobalt alloys
can also be used. Some typical operating parameters of
tape and disc systems are given in Table I.

TABLE I

	Bits per inch	Tracks per inch	Data Rate	Capacity
Tape	100-500	10-20	100-200 KHz	$1-5 \times 10^7$ per tape
Disc	2000-4000	100-200	2-8 MHz	$1-3 \times 10^8$ per pack

Another component of this type of system is the read-
record head and this is usually a toroid of soft magnetic
ferrite with a small non-magnetic gap maintained close to
the recording medium. With a tape it is normally held in
close contact but with a disc system it is more usual to
mount the magnetic head in an air bearing which maintains
the gap $\sim$ 2.5 μm above the disc surface. A coil wound
around the toroid serves both to read and write by means of
the flux changes encountered or induced across the gap.
Some of the limiting factors encountered have been reviewed
by Dudson (1969).

F.   POTENTIAL APPLICATIONS

1.   Bubble domains

        If a thin plate of a uniaxial magnetic crystal with a
saturation magnetisation of a few hundred gauss is subjected
to a small biasing magnetic field, the stable domain
configuration consists of small cylinders of one polarity in
a matrix of the opposite polarity (Fig.3).   Such domains
were reported in orthoferrites by Sherwood et al. (1959) and
a theoretical explanation was proposed by Kooy and Enz (1960)
who observed a similar effect in barium hexaferrite.
Bobeck (1967) showed that these domains could be readily
manipulated, and suggested their application to a number of
logic and storage devices.   The theory of the stability of
these domains has been refined by Thiele (1970) and improved
ways of using these domains in devices have been proposed
(Bobeck et al. 1969).

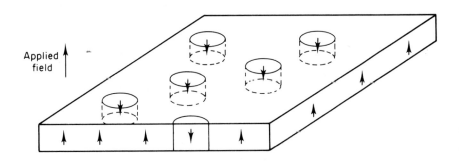

*Fig.3.   Bubble domains in a plate
of uniaxial magnetic material.*

        Cylindrical domains exist as a result of a balance
between the applied bias field plus the domain wall energy,

which tend to collapse the bubble, and the magnetostatic
energy which tends to expand it.   As the applied field is
increased the diameter of the bubble shrinks to a critical
value, at which point it vanishes.   If the applied field is
reduced the bubbles run out into strip domains.   Within
these limits the bubbles can be moved by a small magnetic
field in the plane of the plate, and can be restricted to a
particular set of matrix points (or logical pathways) by a
suitable pattern of permalloy keepers laid down on the
surface.

The size of the domain and the optimum thickness of
the plate are of the same order and are fixed by the magnetic
parameters of the host material.   For most rare-earth orthoferrites the typical domain size is $\sim$ 100 µm but is reduced
by forming solid solutions of $SmFeO_3$ with other rare-earth
orthoferrites (Van Uitert et al. 1970).   The typical domain
size for the barium ferrite family is <1 µm, although this
can be increased by substitution of Al for Fe.   The rare-
earth garnets are cubic and so should not support bubble
domains, but certain planes of flux grown mixed garnets have
a uniaxial character and will support domains in the range
5-10 µm diameter (Van Uitert et al. 1970).   Uniaxial magnetic character can also be induced in a cubic garnet by
strain, and this method has been successfully used to
support bubble domains of a similar size in epitaxial layers
of gallium doped YIG (Mee et al. 1971).

The two parameters of greatest importance for device
applications are domain size, which controls the packing
density, and domain mobility, which controls the data rate
achievable.   So far, it has not been possible to optimise
these two parameters simultaneously.   The problems to be

solved in this area are the cheap reliable production of
uniform slices about 2-3cm square, and the engineering of
an efficient read out system, since the energy stored in
each bubble is very small.

2. Magneto-optic effects

The Faraday effect is large in the visible region in
some materials and could be used as the basis for a large
fast computer store accessed by laser beams (Hunt, 1969;
Eschenfelder, 1970).   A suitable material must have a large
Faraday (or Kerr) effect, a Curie point (or compensation
point) above room temperature and an absorption at a wavelength high enough for the beam to heat the material above
its magnetic transition point but not so high that it unduly
attenuates the read-out beam.   The thermal conductivity
should also be low to keep the resolution high.   EuO has
excellent properties except its Curie point (69°K), and
would require cryogenic operation; alternatively gadolinium
iron garnet, doped to raise the compensation temperature
above room temperature, might be suitable.

Another possible application of the magneto-optic
effect is as an amplitude modulator for light beam based
communication systems, which is feasible at infra-red wavelengths using YIG crystals.   Some attention is now being
paid to $FeBO_3$ which has the advantage that it is transparent
in the green region of the spectrum, but birefringence is a
disadvantage.

3. Spin wave effects

In magnetic crystals with a low spin wave linewidth
the propagation and interaction of magnetostatic,

magnetoelastic and spin waves can be controlled and used as
the basis of delay lines, band stop filters and pulse
compression filters.   Some of these applications have
reached the prototype stage, but such devices are more
likely to be constructed using surface elastic waves in
piezoelectric materials.

IV.   PREPARATION

Nearly all the magnetic oxides in commercial use are
produced in a polycrystalline form by sintering.   A small
amount of single crystal YIG is used but single crystals are
used mainly as vehicles for studying the atomic processes
contributing to the observed magnetic effects.   Recently
effort has been directed to the growth of thin epitaxial
films on non-magnetic substrates, since such a configuration
is of research interest and potential commercial importance.

A.   POLYCRYSTALLINE CERAMICS

Important parameters which must be controlled in
sintered magnetic ceramics are porosity, grain size, and the
homogeneous distribution of the transition metals in the
required valence state.   Variations in these parameters
affects the resistivity, permeability, coercivity, and
resonance line width of the material.

The classical method of preparation of polycrystalline
ferrites involves ball milling the oxides for at least 24
hours, presintering the mix at $\sim$ 1000°C to obtain partial
reaction of the oxides, and further milling before pressing
into the required shape and subjection to the final firing
process.   These steps are critical and the exact details
are usually proprietary.

A different procedure used to ensure intimate mixing
and reactivity, without grinding, is co-precipitation of the
metal carbonates or oxalates followed by thermal decomposition at a relatively low temperature.   Alternatively, the
metal nitrates can be dissolved in alcohol, the solution
atomised with oxygen and ignited.   The product consists of
very fine particles of fully reacted spinel ferrite which
can then be sintered into a required form.

Attempts to provide higher density ferrites with
controlled grain sizes have shown that the normal sintering
of a compressed compact can sometimes be improved by uniaxial
hot pressing in a die at a pressure of $\sim$ 750 bars during the
firing cycle.   A recent development is isostatic hot
pressing in which the material is placed in a thin walled
metal can within an internally heated autoclave with water
cooled walls.   During the firing process the autoclave is
filled with an inert gas to a pressure of $\sim$ 1 k-bar, which
collapses the can and isostatically compresses the sample.

With the highly anisotropic barium hexaferrite
orientated ceramics can be produced by making a slurry of
fully reacted particles and then pressing (with removal of
the liquid) in the presence of a strong DC magnetic field.
Cubic ferrites can be similarly made as textured ceramics
with a low resonance line-width and reduced microwave losses.

B.   SINGLE CRYSTALS

Most magnetic oxides have been grown as single crystals
ranging from mms to cms in size.   The methods available for
the growth of single crystals have been reviewed by Harrison
(1964), Holtzberg (1970), and Kriessman and Goldberg (1963).
The four main methods of growth are crystallisation from the

melt, flux or hydrothermal solution, and from the vapour.

1. Melt growth

Melt growth techniques have not been widely used for
magnetic oxides because the material is required to melt
congruently with a minimum of dissociation and evaporation.
Most magnetic oxides do not behave in this way. Transition
metal oxides and spinel ferrites have been grown by the
Verneuil method, but the products were of very poor crystal
quality. Better quality crystals of magnetite, and also
ferrites containing zinc and cobalt were obtained by Horn
(1961) using the Czochralski technique in a controlled
atmosphere. The best quality magnetite crystals were grown
by Smiltens (1952) using the Bridgman-Stockbarger technique.
$YFeO_3$ has been grown by a modified Bridgman technique in
approximately one atmosphere of oxygen (Blank et al. 1971).
The floating zone technique with arc-image heating has also
been used to grow a range of Ni-Fe ferrites of reasonable
quality in a low oxygen controlled atmosphere (Weaver et al.
1969), and Ni-Zn ferrites have been grown under an oxygen
pressure of 30 atm. (Akashi et al. 1969).

2. Flux growth

The majority of magnetic oxide single crystals have
been grown by some form of flux technique. For flux growth
an inorganic salt is required in which the oxide is soluble
and from which it crystallises as the stable phase. The
ideal flux should have a low melting point, a wide liquid
range, low volatility, low viscosity, low toxicity, and high
purity. The required crystal should not form solid solutions with the flux and its solubility should be $\sim$ 25 wt%

and follow a normal temperature dependent curve. In addition, the flux should not attack the crucible (Pt) and should readily dissolve below $100^{\circ}C$ in some solvent which does not attack the crystals.

In a typical flux run a homogeneous solution is formed, usually in the region of $1000\text{-}1350^{\circ}C$, after which crystallisation occurs by spontaneous nucleation during cooling at rates between 1° and $10^{\circ}C$ per hour. When the crystallisation process is complete in the simplest case the flux is cooled more rapidly to room temperature and dissolved away.

Three main types of flux which have been used are sodium ferrite (which can be regarded as growth from molten sodium carbonate with an excess of Fe_2O_3 present) $BaO\text{-}B_2O_3$ which has the advantages of low volatility, low toxicity, and a better compatibility with platinum than the third group, which is based on PbO, either alone, or with the addition of varying amounts of PbF_2 and B_2O_3 (Kriessman and Goldberg, 1963; Laudise, 1963; Laurent, 1969; Van Uitert et al. 1970).

Growth by slow cooling has a number of disadvantages. The spontaneously nucleated crystals have a dendritic core which considerably reduces the size of useful material. Other problems arise from the volatility of lead salts and the slow process of dissolution in nitric acid required to separate the product from the solidified flux. Various methods to solve these problems have been attempted. Efforts to control nucleation have been made by growth on a seed at constant temperature (Laudise et al. 1962). Alternative solutions include the introduction of a seed crystal into the melt at the temperature of supersaturation followed by further slow cooling (Timofeeva, 1968) or the dipping of a seed crystal into the saturated melt and slow withdrawal, as in the Czochralski technique (Kestigian, 1967).

The problem of evaporation is a serious one, particularly when doped crystals are being grown, as a change in the PbO/PbF_2 ratio in the melt causes a change in the concentration of dopant in the crystal. Attempts to solve this problem include the use of a completely sealed system (Nielsen, 1967) and the use of excess oxygen (Makram et al. 1968).

Flux removal can be effected by pouring off the liquid before it has solidified, but the thermal shock to the crystals is severe. An alternative method drains the crucible from the base and allows the crystals to be furnace cooled (Grodkiewicz et al. 1967).

Despite these efforts slow cooling remains the most successful technique for the growth of large good quality magnetic garnet crystals.

3. Hydrothermal growth

The principles and techniques of hydrothermal growth have been reviewed by Ballman and Laudise (1963). The method is based on the fact that many oxides are soluble in an alkaline supercritical solution to the extent of a few weight per cent and this is sufficient, over a period of days, to transport the desired compound from a hotter ambient zone to a cooler growth zone where it can crystallise out on a seed crystal. This is a low temperature process (< 500°C) and it is often possible to produce high quality crystals, although there may be some incorporation of hydroxyl ions into the crystal lattice. Although small

crystals of ferrite have been produced by the hydrothermal
method it has been most successfully applied to the growth
of YIG (Kolb et al. 1967) and rare-earth orthoferrites (Kolb
et al. 1968) from very concentrated solutions of NaOH or KOH
at temperatures of $\sim$ 400°C and pressures of $\sim$ 500 bars.
Very similar conditions produce good quality gallium doped
Eu/Er iron garnets for bubble domain applications.

4. Vapour growth

Growth by the transport of volatile species through
the vapour phase can be used to produce small, but generally
good quality crystals of many ferrites (Schafer, 1964).
The procedure involves transport from the hot end ($\sim$ 1000°C)
to the cooler end ($\sim$ 800°C) of a sealed silica ampoule
containing a small amount of HCl or Cl_2.   Interesting
variations have been reported such as the modification of
the Piper-Polich technique to grow nickel ferrite by Kleinert
and Schmidt (1968) and the oscillating temperature method of
Scholz and Kluckow (1967).

C.   GROWTH OF THIN LAYERS

Interest exists in the production of thin layers of
ferrites on non-magnetic substrates with the object of
gaining some benefits of planar technology in the microwave
and memory fields, and also with the aim of developing
completely new devices using magneto-optic effects and
bubble domains.

1. Polycrystalline films

A polycrystalline product is not as restrictive in
substrate choice as an epitaxial single crystal layer

(Pulliam, 1967).   Glass, vycor, fused silica, and alumina
ceramics have been used as substrates.   Growth methods
employed include spraying suspensions or solutions of
reactants onto a heated substrate, or coating the substrate
with a layer of reactants by evaporating the solvent from a
suitable solution, followed by a heating cycle to promote
reaction and recrystallisation.   Sputtering techniques,
using either the requisite metals followed by oxidation, or
ceramic ferrites and garnets directly, have also been
employed.   Ferrite films have also been deposited by a
vacuum arc discharge method (Nave and Yamanaka, 1970).

The two methods which offer the best possibilities for
practical application to the production of microwave devices
are DC arc-plasma spraying followed by an annealing cycle
(Harris et al. 1970), and a chemical transport deposition
technique developed by Braginski and Buck (1969).   In this
method the ferrite was transported across a narrow gap as
chlorides and deposited on a variety of substrates at temperatures above 1000°C.   In both these cases Mn-Mg ferrites
suitable for use in microstrip devices were produced.

2. Epitaxial films

Extensive research has been devoted to the growth of
thin layers of magnetic oxides on single crystal substrates,
e.g. magnesium oxide, $MgAl_2O_4$ spinel, and non-magnetic
garnets, where there is a sufficiently close structural
relationship to promote epitaxial growth.   Most work has
used chemical vapour transport techniques, but epitaxial
films of ferrites, garnets and orthoferrites have been prepared by rf sputtering techniques.   There have also been
reports of the successful growth of epitaxial films from a

flux, such as spinel ferrites on MgO using Na_2CO_3 as the
flux (Gambino, 1967), YIG on gadolinium gallium garnet
from a lead borate flux (Linares, 1968) and orthoferrites
on themselves from $PbO:B_2O_3$ mixtures (Schick and Nielsen,
1971).

The growth of epitaxial layers by the chemical transport method has been reviewed by Mee and Pulliam (1969) who
have further developed the original work of Cech and
Alessendrini (1939) on the growth of FeO, NiO and CoO layers
on MgO. Some workers have continued to use the simple
close-spaced arrangement of Cech and Alessendrini, but the
best controlled growth has been achieved in open flow
systems using either concentric tube reactors (Stein, 1970;
Marshall, 1971) or T-shaped reactors (Mee, 1967).

Provided that the general structure is sufficiently
similar (e.g. MgO for spinel ferrites, garnets for YIG) the
choice of substrate is not critical for obtaining initial
epitaxial growth. If layers thicker than 1 or 2 μm are
required it is necessary to obtain a good match of thermal
expansion between layer and substrate, otherwise the layer
will be subjected to severe stresses and cracking. This is
of particular importance when gallium-doped YIG layers are
being grown for use as a magnetic bubble host since the
required anisotropy is strongly affected by strain, and can
be controlled to a certain extent by substrate selection
(Mee et al. 1971). With suitable development the chemical
vapour deposition technique could provide large scale
economic production of epitaxial layers for bubble domain
devices.

REFERENCES

Akashi, T., Matumi, K., Okada, T. and Mizutani, T. (1969)
I.E.E.E. Trans. Magnetics Mag-5, 285.

Ballman, A.A. and Laudise, R.A. (1963) "The Art and Science
of Growing Crystals" (J.J. Gilman, ed.) Wiley, New York.

Blank, S.L., Schick, L.K. and Nielsen, J.W. (1971) J. Appl.
Phys. 42, 1556.

Bobeck, A.H. (1967) Bell Syst. Tech. Journ. 46, 1901.

Bobeck, A.H., Fischer, R.F., Perneski, A.J., Remeika, J.P.
and Van Uitert, L.G. (1969) I.E.E.E. Trans. Magnetics
Mag-5, 544.

Bobeck, A.H., Van Uitert, L.G., Abrahams, S.C., Barns, R.L.,
Grodkiewicz, W.H., Sherwood, R.C., Schmidt, P.H.,
Smith, D.M. and Walters, E.M. (1970) Appl. Phys. Lett.
17, 131.

Bozorth, R.M. (1969) I.E.E.E. Trans. Magnetics Mag-5, 692.

Braginski, A.I. and Buck, D.C. (1969) I.E.E.E. Trans. Mag-5,
924.

Brailsford, F. (1966) "Physical Principles of Magnetism",
D. Van Nostrand, London.

Cech, R.E. and Alessendrini, E.I. (1959) Trans. A.S.M.
50, 150.

Dudson, M.F. (1969) Radio and Electronic Engineer 38, 227.

Eschenfelder, A.H. (1970) J. Appl. Phys. 41, 1372.

Gambino, R.J. (1967) J. Appl. Phys. 38, 1129.

Geller, S. (1956) J. Chem. Phys. 24, 1236.

Grodkiewicz, W.H., Dearborn, E.F. and Van Uitert, L.G. (1967)
"Crystal Growth" (H.S. Peiser, ed.) Pergamon Press,
Oxford p.441.

Harris, D.H., Janowiecki, R.J. Semler, C.E., Willson, M.C.
and Cheng, J.T. (1970) J. Appl. Phys. 41, 1348.

Harrison, F.W. (1964) Proc. Brit. Ceram. Soc. 2, 91.

Harvey, A.F. (1963) "Microwave Engineering", Academic Press, New York.

Holtzberg, F. (1970) J. Appl. Phys. 41, 1283.

Horn, F.M. (1961) J. Appl. Phys. 32, 900.

Hudson, A.S. (1970) The Marconi Review 21.

Hunt, R.P. (1969) I.E.E.E. Trans. Magnetics Mag-5, 700.

Jacobs, I.S. (1969) J. Appl. Phys. 40, 917.

Kestigian, M. (1967) J. Amer. Ceram. Soc. 50, 165.

Kleinert, P. and Schmidt, P. (1968) Zeit. Chem. 8, 395.

Kolb, E.D., Wood, D.L., Spencer, E.G. and Laudise, R.A. (1967) J. Appl. Phys. 38, 1027.

Kolb, E.D., Wood, D.L. and Laudise, R.A. (1968) J. Appl. Phys. 39, 1362.

Kooy, C. and Enz, U. (1960) Philips Research Reports 15, 7.

Kriessman, C.J. and Goldberg, N. (1963) "Magnetism" Vol.3 (G.T. Rado and H. Suhl, eds.) Academic Press, New York.

Laudise, R.A. (1967) "The Art and Science of Growing Crystals" (J.J. Gilman, ed.) Wiley, New York.

Laudise, R.A., Linares, R.C. and Dearborn, E.F. (1962) J. Appl. Phys. 33S, 1362.

Laurent, Y. (1969) Rev. Chimie Minerale 6, 1145.

Lax, B. and Button, K.J. (1962) "Microwave Ferrites and Ferrimagnetics" McGraw Hill, New York.

Linares, R.C. (1968) J. Cryst. Growth 3, 4, 443.

Makram, H., Touron, L. and Loriers, J. (1968) J. Cryst. Growth 3, 4, 452.

Marshall, D.J. (1971) J. Cryst. Growth 9, 305.

Mee, J.E. (1967) I.E.E.E. Trans. Magnetics Mag-3, 190.

Mee, J.E., Pulliam, G.R., Archer, J.L. and Besser, P.J. (1969) I.E.E.E. Trans. Magnetics Mag-5, 717.

Mee, J.E., Pulliam, G.R., Heinz, D.M., Owens, J.M. and
   Besser, P.J. (1971) Appl. Phys. Lett. 18, 60.

Nave, M. and Yamanaka, S. (1970) Japan. J. Appl. Phys. 9,
   293.

Néel, L. (1948) Ann. Phys. 3, 137.

Nielsen, J.W., Lepore, D.A. and Leo, D.C. (1967) "Crystal
   Growth" (H.S. Peiser, ed.) Pergamon Press, Oxford, P.457.

Pulliam, G.R. (1967) J. Appl. Phys. 38, 1120.

Rado, G.T. and Suhl, H. (1963) "Magnetism" Academic Press,
   New York and London.

Richard, B.W. (1970) I.E.E.E. Trans. Magnetics Mag-6, 791.

Schick, L.K. and Nielsen, J.W. (1971) J. Appl. Phys. 42,
   1554.

Scholz, H. and Kluckow, R. (1967) "Crystal Growth"
   (H.S. Peiser, ed.) Pergamon Press, Oxford, p.475.

Schäfer, H. (1964) "Chemical Transport Reactions" Academic
   Press, New York.

Sherwood, R.C., Remeika, J.P. and Williams, H.J. (1959)
   J. Appl. Phys. 30, 217.

Smiltens, J. (1952) J. Chem. Phys. 20, 990.

Smit, J. and Wijn, H.P.J. (1959) "Ferrites" J. Wiley and
   Son, New York.

Stein, B.F. (1970) J. Appl. Phys. 41, 1262.

Thiele, A.A. (1970) J. Appl. Phys. 41, 1139.

Timofeeva, V.A. (1968) J. Cryst. Growth 3, 4, 496.

Treves, D. (1965) J. Appl. Phys. 36, 1033.

Van Uitert, L.G. Sherwood, R.C., Bonner, W.A.,
   Grodkiewicz, W.H., Pictroski, L., and Zydzik, G.J.
   (1970a) Mat. Res. Bull. 5, 153.

Van Uitert, L.G., Smith, D.H., Bonner, W.A., Grodkiewicz, W.H.
   and Zydzik, G.J. (1970b) Mat. Res. Bull. 5, 455.

Van Uitert, L.G., Bonner, W.A., Grodkiewicz, W.H.,
  Pictroski, L. and Zydzik, G.J. (1970c) Mat. Res. Bull. $\underline{5}$,
  825.

Weaver, E.A., Merchant, M.D. and Poplawsky, R.P. (1969)
  J. Amer. Ceram. Soc. $\underline{52}$, 214.

White, R.L. (1969) J. Appl. Phys. $\underline{40}$, 1061.

White, R.M. (1970) Proc. I.E.E.E. $\underline{58}$, 1238.

OXIDES AS SUBSTRATES IN MICROELECTRONICS

J.D. FILBY

Royal Radar Establishment,
Malvern, Worcestershire

I.   INTRODUCTION

Oxide materials play an extremely important part in microelectronics both in the fabrication of devices and as an integral part of their structure. For example, in the manufacture of silicon devices extensive use is made of silicon oxides as masks to delineate the areas into which electrically active impurities are diffused. The diffusion source is normally an oxide-glass deposited on the surface

of the device structure, Fig.1a.   Oxides are also used in
certain devices to isolate metallic contacts (gates) from
the silicon.   By applying a voltage to these gates the
amount of current passing through the silicon below the
oxide can be controlled, Fig.1b.   It is fortunate that very
pure silicon oxides are relatively easily formed on a silicon
surface, are stable at high temperatures, impervious to the
standard dopants, relatively easy to remove by etching techniques, and are basically electrically inert.   This review
is concerned with uses of oxides in microelectronics as substrates.   Insulating substrates have been an integral part
of microelectronics since the development of thin film
microcircuits.   However, it is only since the introduction
of single crystal substrates for the growth of epitaxial
semiconductor layers that the detailed structure and chemical
behaviour of the substrate has contributed significantly to
the properties of the deposited film.   Previously the
dependence of the film characteristics on the amorphous substrates used in thick and thin film technologies, was mainly
limited to topographical and adhesion considerations.   With
the increasing circuit complexity and the development of new
technologies (e.g. microwave hybrids) the chemical purity,
homogeneity and thermal conductivity of glass and ceramic
substrates are becoming more important.

II.   SUBSTRATES FOR SINGLE CRYSTAL FILMS
A.   INTRODUCTION
       Since the resistivity of silicon is generally low,
special techniques have been devised to electrically isolate
adjacent circuits.   The most common approach is to surround
each circuit by a back-biased junction which creates a

narrow isolation region, Fig.1c.   Unfortunately this
introduces parasitic capacitance effects which limit the
operational speed of the circuit.   To overcome this,
dielectric isolation techniques have been developed in which
each circuit is surrounded by an oxide barrier, Fig.1d.
Because of thermal expansion differences between the oxide
and silicon the oxide cannot be very thick and electrical
isolation is not perfect.   Complete isolation can be
obtained by depositing a single crystal semiconductor film
on an insulating substrate and then using etching techniques
to produce discrete islands, Fig.1e.

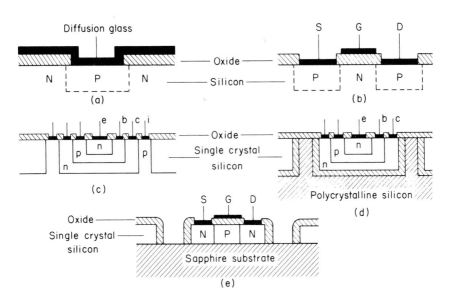

Fig.1.   a) Oxide mask and diffusion glass.

b) Oxide used as gate (G) dielectric.

c) Junction isolation, i negative w.r.t. c.

d) Dielectric isolation.

e) Insulating substrate, silicon on sapphire.

TABLE I

Type	Softening Point (°C)	Density (g/cm³)	Thermal Conductivity at 25°C (cal/cm/sec/°C)	Dielectric Constant (1MHz at 25°C)	Loss Tangent (1MHz at 25°C)
Soda lime glass [a]	696	2.47	0.0023	6.9	0.01
Alkali borosilicate (Pyrex) [a]	820	2.23	0.0027	4.6	0.0062
Barium alumino-silicate (alkali free, 7059 glass) [a]	872	2.76	0.0030 [b]	5.8	0.0011
Polished fused silica [a]	1580	2.20	0.0035	3.7	0.00002
Glazed 96% alumina [b]	740 for glaze	3.7 (bulk)	0.08 (bulk)	8.9	0.004
Glazed 96% beryllia [b]	740 for glaze	3.0 (bulk)	0.5 (bulk)	6.0 (bulk)	-
94% Alumina +CaO+SiO$_2$ [a]	1500	3.58	0.073	8.9	0.0018
96% Alumina +MgO+SiO$_2$ [a]	1550	3.70	0.084	9.3	0.0028
99% Alumina	-	3.90	-	9.95	0.0007
99.5% Beryllia [a]	1600	2.88	0.55	6.4	0.0006
Single crystal sapphire [b]	2040 MPT	3.98	0.08	10.25(10GHz)[c]	0.0006 (10GHz)[c]

a - Macavoy and Halaby (1964)　　　b - Schwartz and Berry (1964)　　　c - Bosnell (1971)

Single crystal films are usually grown by epitaxial techniques on single crystal substrates although amorphous substrates have been studied in efforts to increase the availability of suitable substrate materials. Amorphous substrates would also be cheaper and not subject to the size limitations of single crystals.

B.   SINGLE CRYSTAL SUBSTRATES

1. Introduction

In order to use epitaxial techniques there must be some possibility of an atomic fit at the substrate/film interface. An indication of compatibility may be seen by comparing the lattice parameters, although the relationship is not always obvious if the crystal structures are markedly different. Since the epitaxial deposition process occurs at a high temperature, $\sim 1000^{\circ}C$, the film and substrate must also have similar thermal expansion coefficients. In addition, the substrate must be chemically stable at the deposition temperature and not subject to attack by any of the reactants or products of the deposition process. It is also an advantage if the mechanical properties of the substrate allow it to be polished to give a reproducible damage free surface. Typical properties of substrate materials and the deposited semiconductors are given in Table I. Crystal growth and electrical properties are not extensively discussed as these are noted in another chapter, and in the reviews of Allison et al. (1969) and Cullen (1971).

2. Structural investigations

Any structural discontinuity in the substrate, such as a low angle grain boundary, is reproduced in the deposited

epitaxial film.   Most of the early sapphire substrates were
cut from crystals grown by the Verneuil process and contained such boundaries in significant numbers.   They can be
readily revealed by etching techniques (Scheuplein and Gibbs,
1960).   The Verneuil crystals were highly strained and
often shattered when being cut, polished, or heated to the
semiconductor deposition temperature.   To overcome these
material limitations, sapphire crystals grown by the more
controllable Czochralski technique were employed.
Czochralski-grown crystals often contained no low angle
grain boundaries and their dislocation content is 10^2-10^3/cm^2,
compared to 10^5-10^6 for the Verneuil material.   The quality
of Verneuil material has subsequently been improved and
selection of crystals free from low angle boundaries, by
viewing between crossed polaroids, allows substrates as
reliable as the more expensive Czochralski material to be
obtained.

Since sapphire belongs to the rhombohedral class the
possibility of a reasonable atomic match with silicon
(diamond cubic structure) is remote.   The cubic magnesium
aluminate spinel has therefore been investigated as a
possible substrate material.   Unlike sapphire, this spinel
system does not have a unique composition but exists over a
wide compositional range.   Early spinel substrates were cut
from Verneuil crystals of composition $MgO:3.3\ Al_2O_3$, but
contained numerous low angle boundaries and were structurally
inferior to the Verneuil sapphire (Manasevit and Forbes,
1966).   They were also thermally unstable, and heating to
the deposition temperature resulted in cracking and the
exsolution of excess alumina.   Further, transition metal
impurities, added to improve the spinel growth characteristics,

adversely affected the perfection and electrical properties
of the epitaxial layers (Heiman and Robinson, 1968).
Because of these deficiencies, stoichiometric spinel has
been grown by the Verneuil flame fusion (Arlett and Robbins,
1967), the Czochralski (Cockayne and Chesswas, 1967;
Grabmaier and Watson, 1968) and flux methods (Wang and
MacFarlane, 1968).   Stoichiometric material does not
exhibit the thermal instability of the alumina-rich material.
Flame fusion material is generally too strained to be useful,
whereas Czochralski material is satisfactory and of similar
quality to the Czochralski sapphire.   However, due to
relatively high cost and difficulties experienced in
obtaining a satisfactory surface finish, it has not been
extensively used.   Wang and MacFarlane (1968) have shown
that the dislocation density in flux grown material is as
low as $50/cm^2$ (compared to 10^5 and $10^3/cm^2$ for Verneuil and
Czochralski material respectively): it is also free from
any subgrain structure.   Flux grown material, however, is
too irregular in size and shape to be commercially attractive
as a substrate material.   In view of this and the difficulties in polishing stoichiometric spinel, Wang (1969) has
carried out a detailed study of the structural perfection of
a range of alumina rich spinels ($MgO:1-3\ Al_2O_3$) grown by the
Verneuil technique.   He found that the ease of growth and
thermal stability are functions of crystal composition.
Crystals containing up to $MgO:2.5\ Al_2O_3$ are thermally stable,
and heat treatments up to $1200^{\circ}C$ produce no cracking or
exsolution of the excess alumina.   Growth is more difficult
the lower the alumina content, and only small stoichiometric
crystals are possible.   Wang's crystals are in general more
perfect than the commercial flame fusion material, but Berg-

Barrett X-ray topographical studies reveal some subgrain
structure and they are inferior to the best flux grown
material.   Wang concludes that crystals in the composition
range $MgO:1.5-2.5$ Al_2O_3 are suitable as substrates for
silicon epitaxy, also confirmed by later work (Cullen et al.
1969; Zainginger and Wang, 1970).

For various reasons none of the other substrate
materials listed in Table I have been extensively investigated.   The high thermal expansion coefficient for single
crystal quartz, compared with silicon, results in the quartz
fracturing on cooling from the deposition temperature if the
silicon layer is thicker than 4μ (Joyce et al. 1965).
Manasevit and Forbes (1966) found that magnesium oxide substrates were not as good as sapphire because of the chemical
reactivity of the halides produced during the deposition
process.   Filby and Nielsen (1967) investigated the use of
yttrium aluminium garnet, calcium tungstate and stannic
oxide as substrate materials and found that they interacted
with the depositing silicon above $1000^{o}C$.   Thorium dioxide
(ThO_2), zircon ($ZrSiO_4$) and lanthanum aluminate ($LaAlO_3$)
(Chu et al. 1965) are only available in small platelet form
and have received little attention.   Because of its high
thermal conductivity, beryllium oxide (BeO) is a very
attractive substrate material.   Unfortunately, it is only
available as small flux grown crystals a millimeter or so
across.   Manasevit et al. (1966) have shown that silicon
can be successfully grown on the natural faces of these
small crystals.

3.   Surface preparation

The crystallographic and electrical quality of the

epitaxial film is very dependent upon the surface finish of
the substrate. Since nucleation and deposition can take
place preferentially at scratches on poorly polished substrates, it is necessary to have a smooth work-damage-free
surface to obtain reproducible high quality films. After
cutting from the boule, most substrates are lapped flat and
mechanically polished to $\frac{1}{4}\mu$ diamond (sapphire) or $\frac{1}{4}\mu$ alumina
(spinel). At this stage the surface is covered by a work-
damaged layer which has to be removed by an etching
technique. Several acids or molten salts have been
investigated as possible chemical etches, but they have the
disadvantage that a residue can be left on the surface which
may contaminate the deposited layer. Several alternative
gaseous etches have been investigated. For both sapphire,
(Robinson and Mueller, 1966) and spinel, (Wang et al. 1969)
annealing in hydrogen can produce smooth damage-free surfaces, although the exact etching mechanism is uncertain.
Dumin (1967a) has reported that pre-firing in the presence
of helium or oxygen does not remove the work-damage layer.

 Hydrogen annealing of spinel is not as successful as
for sapphire. Wang (1969) reports that although a 1 hr.
anneal at $1200^{\circ}C$ will remove the amorphous layer from (111)
spinel it does not remove the scratch marks. If the
annealing period is increased, etch pits may be produced.
It is therefore necessary to obtain a high quality mechanical, or chemical, finish before using the hydrogen in situ
anneal as a final pre-deposition step.

 Manasevit and Morritz (1967) have reported the
successful use of sulphur fluorides as polishing etchants for
sapphire. Sulphur tetrafluoride diluted in hydrogen or
helium produced a highly polished scratch-free surface on

substrates heated above 1450°C.   The hexafluoride was also
successful when diluted in helium, but when hydrogen was
used as the carrier gas little attack was observed under
otherwise identical conditions.   The exact conditions for
satisfactory etching were found to be orientation dependent.
The results obtained with the sulphur fluoride were more
reproducible than those for anhydrous hydrogen chloride and
the etching rate considerably faster.

Silicon itself can be used as a polish and this has
been extensively investigated using vacuum deposition technology, since it offers the possibility of an in situ etch.
Reynolds and Elliott (1967) found that for substrate temperatures in the range 950-1100°C there existed a critical
flux rate of impinging silicon vapour, below which no
deposition took place and the sapphire surface was etched.
Naber and O'Neal (1968) have shown that at 1300°C the
etching rate is a linear function of the silicon impingement
rate and is orientation dependent;   e.g. for a given flux,
the (1$\bar{1}$02) sapphire plane etches faster than the basal plane.
If too high an etch rate is used, etch pits are formed and
the surface roughness increases.   Polishing using silicon
is also possible in a chemical vapour deposition system.
Filby (1966) has reported that a very diluted concentration
of silane in hydrogen can polish sapphire substrates at
temperatures near the critical condensation temperature.
Dumin and Robinson (1968) have subsequently reported that
epitaxial films grown on SiH_4-H_2 etched substrates are
heavily contaminated with aluminium.

Hart et al. (1967) concluded that the electrical properties of films grown at 1095°C were not dependent on the
method of substrate preparation or the perfection of the

sapphire. They obtained similar mobilities on mechanically and/or chemically polished Verneuil or Czochralski material provided that the quality of the polish was satisfactory. Hydrogen pre-firing was found to be beneficial if the films were deposited at lower temperatures ($1000^{\circ}C$) but otherwise it only improved the reproducibility.

4. Chemical stability during deposition

Substrate attack during deposition can have two major detrimental effects. It can result in the incorporation of impurities in the deposited film and, if non-uniform, can produce films of poor crystallographic quality. This is because the silicon/sapphire orientation relationship varies with the substrate orientation and, if the surface is not a single plane, multi-orientated films may result.

Since aluminium is an electrically active impurity in silicon it is particularly important that no interaction occurs during deposition to release aluminium from the sapphire or spinel substrates. Early work indicated that films grown on sapphire from silane, SiH_4, were less heavily contaminated, and crystallographically more perfect, than those grown from trichlorosilane, $SiHCl_3$, (Bicknell et al. 1966) or silicon tetrachloride, $SiCl_4$ (Manasevit et al. 1965). Both groups of workers observed significant differences in the nucleation and growth behaviour between the hydride and halide deposited films. In the latter case there were indications that the substrate had been more severely attacked and, since the nucleation density was lower, that the attack had taken place over a longer period.

The exact mechanism by which this attack takes place has not been established, mainly because of the uncertainty that exists over the decomposition kinetics for the halides on a clean sapphire surface (Manasevit et al. 1965). It is almost certainly associated with the production of hydrogen chloride during the deposition reaction. Bicknell et al. (1966) have suggested a reaction of the type

$$Al_2O_3(s) + 6HCl(g) \rightarrow 2AlCl_3(g) + 3H_2O(g)$$

which can proceed rapidly above 850°C if the gaseous reaction products can escape. This type of attack would not be possible in the silane case since no halides are involved. Silicon will also etch exposed sapphire surfaces at the deposition temperature and it is necessary, even when using silane or evaporation techniques, to minimise the time during which a free sapphire surface exists. This may be done by lowering the deposition temperature but with adverse effects on the crystalline perfection and hence the electrical properties of the deposited film. Further, not all the contamination may originate from the front surface of the substrate (Hart et al. 1967). The reverse side can act as a continuous source, since it is difficult to prevent a slow silicon attack at this hotter surface. It is true that for sapphire (Dumin 1967a), and to a lesser extent spinel (Wang et al. 1969), it is possible to control the effect of the aluminium contamination by a post-growth oxidation treatment. Such a treatment is only partially successful, and it is better to minimise the contamination initially.

In practice a compromise has to be found by experiment between the contamination level and the crystalline perfection of the deposited film. This is usually done by depositing films over a range of growth temperatures, and subsequently measuring the Hall mobility and carrier

concentration.   The results of Dumin (see Allison et al.
1969) clearly demonstrate that there is an optimum growth
temperature for maximum mobility, and that the carrier
concentration increases with growth temperature.   Both
properties are functions of substrate material and orientation.   In this context spinel has particular advantages as
a substrate for high mobility films because of a lower
contamination level.   The improved stability of spinel
compared to sapphire under the deposition conditions is
emphasised by the fact that Seiter and Zaminer (1965) have
grown satisfactory films using silicon tetrachloride as the
source.

Germanium films have been grown on sapphire (Tramposch,
1966, 1969) and spinel (Dumin 1967b) using close-spaced
vapour transport and hydride vapour deposition techniques
respectively.   No aluminium doping or other indication of
substrate attack was observed, probably due to the slowness
of any interaction at the reduced deposition temperature,
$\sim 700^{\circ}C$.

The successful growth of III-V compounds is not
possible, due to substrate attack, if the conventional
methods involving hydrogen chloride are used (Manasevit and
Simpson, 1969).   The development of techniques using metal-
organic sources, e.g. triethylgallium and trimethylgallium,
has resulted in the successful growth of III-V compounds on
a wide range of substrates (Manasevit, 1968; Manasevit and
Simpson, 1969; Rai-Choudhury, 1969).   Manasevit and
Thorsen (1970) have made a detailed investigation of the
nucleation and growth stages for gallium arsenide deposited
on sapphire but do not report any indications of substrate
attack.

5. Interface structure

The detailed physical and chemical structure of the interface region is uncertain. It is widely accepted that, in the case of sapphire, a highly disordered layer is formed at the interface containing silicon, oxygen and aluminium. The disturbed region can act as an impurity sink during subsequent device processing, producing undesirable electrical effects (Heiman, 1967). Mercier (1970) has used infrared spectroscopy to confirm the presence of Si-Al and Si-O bonds in silane deposited layers, but it was not possible to determine whether these were at the interface. The absence of any stable interface compound is indicated by the fact that if the silicon layers are removed from either sapphire or spinel, the substrate surface appears structurally unchanged when examined by electron diffraction. Transmission electron microscopy studies of the interface have been virtually impossible because of the difficulty in thinning down the oxide without attacking the silicon. Schlötterer and Zaminer (1966) have used an argon ion etching technique to thin down silicon-on-spinel samples, and have been able to study the occurrence of dislocations, stacking faults and twins. The use of ion-beam etching techniques and high voltage microscopy suggests that it should soon be possible to obtain a detailed picture of the interface structure. Electron energy loss analysis may yield information on the chemical species present.

6. Deformation and stress

Since the semiconductor films are always grown at elevated temperatures, and the thermal expansion coefficients of silicon and the oxide substrates differ,

stresses are produced in the combined structure on cooling
to room temperature.   If the stress exceeds the yield
stress of either the film or substrate then it is possible
for fracture to occur, e.g. silicon on quartz.   Before the
stresses reach these proportions they can have detrimental
effects on the electrical properties of the films (Paul and
Pearson, 1955).   The deformation of silicon-on-sapphire
films has been measured by Dumin (1965) and Ang and
Manasevit (1965).   The films are under a compressive stress
of $\sim 5 \times 10^9$ dyn/cm^2.   This is theoretically large enough to
produce an electrical effect, but has not been experimentally
observed.

The compressive stress in silicon films deposited on
spinel is of the same magnitude and has been determined for
alumina rich spinel by Schlötterer (1968) and Robinson and
Dumin (1968), and for stoichiometric spinel by Wang et al.
(1969).   Robinson and Dumin reported the stress to be
anisotropic, in disagreement with the other investigations.
The anisotropic effect may have been due to the exsolution
of α - Al$_2$O$_3$ at dislocations and subgrain boundaries during
processing.

Since the linear thermal expansion coefficient for
gallium arsenide is closer than that of silicon to the
values for sapphire and spinel it would be expected that
the residual stress would be less.   Manasevit and Simpson
(1969) have confirmed this and report that the small
residual stress in the gallium arsenide is compressive.
Similarly, Dumin (1966) has reported the deformation of
germanium-on-spinel films to be less than for silicon on
spinel;   again the result of a closer match in the coefficients of thermal expansion and the lower growth temperature.

7. Orientation relationships

When the film and substrate are structurally similar
the orientation relationships are usually very simple, e.g.
(111)Si $\|$ (111)Spinel.   This is the case for silicon and
gallium arsenide deposited on spinel (Manasevit and Forbes,
1966; Manasevit and Simpson, 1969), thoria, zircon or
lanthanum aluminate (Chu et al. 1965).   If a particular
type of interface bonding is favoured exceptions can occur,
for example, Manasevit and Simpson (1969) have obtained (100)
GaAs on (110) Czochralski spinel.

The crystallographic relationship between two dis-
similar structures such as silicon (diamond cubic) and
sapphire (rhombohedral) is far more complicated.   Manasevit
et al. (1968) have investigated in detail the crystallography
of the silicon/sapphire system for all possible sapphire
orientations.   There are seven single crystal relationships
and the boundaries between these are clearly defined.   The
range of substrate orientations over which four multi-
orientated relationships exist have also been established.
The results give a clear explanation of discrepancies noted
in earlier studies (Filby and Nielsen, 1967).

Since silicon and gallium arsenide are of similar
cubic structure it would be expected that the epitaxial
relationships for GaAs on sapphire would be similar to those
for silicon.   The only one reported, (111)GaAs $\|$ (0001)Al$_2$O$_3$,
is similar, but Manasevit et al. (1968) report that prelim-
inary investigations of other sapphire orientations suggest
that a different set of relationships exists.

It is possible that a similar range of discrete
orientation relationships exists for silicon and gallium
arsenide deposited on single crystal beryllia (Manasevit and

Simpson, 1969).   Unfortunately, only films grown on the
natural faces have been studied, due to the lack of single
crystal material and the hazards involved in machining
(Manasevit et al. 1966; Manasevit, 1968).   The only other
system to be studied in detail is silicon on quartz.   Using
glancing-angle diffraction techniques, Bicknell et al. (1964)
have established that the relative orientation between the
silicon and quartz is invariant and corresponds to {001}Si ||
(0001)Quartz and {010}Si || {01$\bar{1}$0}Quartz.

Several attempts have been made to explain the
occurrence of the observed relationships for the silicon-on-
sapphire system in terms of the atomic arrangement at the
interface (Nolder and Cadoff, 1965; Larssen, 1966) or the
propagation of crystallographic symmetry elements (Bicknell
et al. 1966), but experimental proof of the proposed structures has not been obtained.

C.   POLYCRYSTALLINE AND AMORPHOUS SUBSTRATES

The use of single crystal materials as substrates is
very restrictive, and if techniques could be developed to
produce single crystal films on amorphous or polycrystalline
materials several advantages would be gained.   Substrates
would become available in a wider range of sizes, there
would be no problem of atomic mismatch at the interface, and
the constituents of the material could be adjusted to give a
coefficient of thermal expansion similar to that of the semiconductor.

The only non-single crystal materials to receive
attention as possible substrates for the growth of silicon
single crystal films are amorphous quartz and polycrystalline
alumina.   Since these substrates basically possess no

ordered atomic structure, the deposited nuclei will be
randomly orientated and epitaxial techniques cannot be used.
Substantial single crystal areas can only be obtained if the
number of nuclei can be reduced sufficiently for large areas
to be grown from a single nucleus.   This can be achieved by
using a travelling mask which restricts the nucleation stage
and encourages growth to proceed in the direction of travel,
on the initial nuclei.

The simplest method of reducing the nucleation density
is to grow the film from the liquid rather than vapour phase.
Doo (1964) reported substantial grain growth in polycrystalline silicon films (on alumina) when heated 5 - 15°C
above the melting point.   An alternative way of introducing
a liquid stage is to form a low-melting point alloy with the
silicon.   If silicon is deposited on this alloy the excess
separates out as a thin film (Filby and Nielsen, 1965).
Unfortunately, the liquid alloys used tend to vigorously
etch the quartz substrates, and the silicon deposition rate
and temperature have to be very carefully controlled if surface attack is to be avoided.   The technique has been
adopted by Filby and Nielsen (1966) and Haidinger et al.
(1966) to give large crystallites on pre-selected areas.
Braunstein (1968) has reported a combination of the travelling mask and alloy techniques.   The degree of success from
any of these techniques is limited, and the reproducible
growth of single crystal films on amorphous substrates has
still to be achieved.

III.   SUBSTRATES FOR THIN AND THICK FILM CIRCUITS

A.   INTRODUCTION

Thin film circuits consist of conductors, resistors,

capacitors and, to a lesser degree, inductors deposited on
an insulating substrate by evaporation, sputtering, anodisa-
tion, pyrolytic or chemical deposition techniques (Campbell,
1966).   The various components are delineated by masking
during deposition, or by subsequent selective chemical or
ion beam etching techniques.   The major advantages over
using discrete components are a saving in weight and volume,
and an increase in reliability by elimination of discrete
interconnections.   The film thickness is usually less than
1μ hence the terminology 'thin film' compared with 'thick
films' which are $\sim 40\mu$ thick at the printing stage and 10μ
or so after firing.   Thick film conductors, resistors and
capacitors are printed by squeezing an 'ink' through a
series of nylon or stainless steel screens which are
impervious except where the required circuit has been
delineated by photo-resist techniques.   The printing is
followed by a firing sequence which sinters and stabilises
the film.   The advantage of screen printing is that it is
cheap and, since it allows a thicker film to be deposited
than by any other printing technique, it assures the
continuity of conductors and the freedom from pinholes for
dielectrics deposited on relatively rough, low-cost substrates (Holmes and Corkhill, 1969).

It is worth mentioning that oxides are also involved
in thin films as the major source of capacitor materials,
e.g. silicon monoxide and dioxide, and that oxides form an
integral part of thick film inks.   The inks usually contain
metallic components, a glass frit and an organic vehicle,
the latter being removed during the firing process.   The
glass frit is responsible for reacting with the substrate
during firing to bond the film, and it is also possible with

TABLE II

Material	Crystal Structure	Lattice Parameter (Å)	Melting Point (°C)	Coefficient of Linear Expansion (10^{-6}/°C^{-1})	Single Crystal Availability
Si	Cubic (diamond)	a = 5.431	1430	3.9 (50-1000°C)	
Ge	Cubic (diamond)	a = 5.657	958	5.658	
GaAs	Cubic (zinc blende)	a = 5.653	1280	6.68	
Al_2O_3	Rhombohedral	a = 4.7, c = 12.97 ref. hex. cell	2040	9.5 (20-2000°C)	Flux, Verneuil & Czochralski grown single crystals
$MgO.XAl_2O_3$	Cubic	$MgO.Al_2O_3$ a = 8.11 $MgO.2Al_2O_3$ a = 8.02 $MgO.3Al_2O_3$ a = 7.98	2130 2080 2000	8.8 (20-1200°C) 8.1 (50-1000°C)	Flux, Verneuil & Czochralski grown single crystals
α-SiO_2	Hexagonal	a = 4.9, c = 5.39	1710	18.0 (0-1000°C) (⊥r to C axis)	Hydrothermally grown single crystals
BeO	Hexagonal	a = 2.7, c = 4.39	2530	9.0 (0-1200°C)	Flux grown crystals
ThO_2	Cubic	a = 5.60	3300	9.4 (0.1200°C)	Single crystal platelets
$LaAlO_3$	Rhombohedral <435°C Cubic >435°C	a = 3.79, c = 90°5' a = 3.81	2000 - 2200	12.0 (25-650°C)	Single crystal platelets
$ZrSiO_4$	Tetragonal	a = 5.58, c = 5.93	2390 - 2550	5.5 (0-1200°C)	Single crystal platelets
MgO	Cubic	a = 4.20	2850	14.4 (20-1200°C)	Arc-melted single crystals

some inks for the glass to form a protective surface, thus
eliminating the need for any protective glazing.   During
the firing cycle the metallic components may be oxidised;
it is the degree of oxidisation, in conjunction with the
ratio of metallic to glass constituents, that determines the
final resistivity of the film (Kelemen, 1970).

Thick film substrates have to be flat, relatively
smooth but not glazed, and capable of withstanding processing
temperatures in excess of $1000^{\circ}C$ (Szekely, 1969).   Glass
substrates are not suitable since they soften below the
processing temperature, Table II.   A ceramic material is
therefore necessary, and alumina is the most widely used
since it is readily available, cheap and has good thermal
and electrical properties.   Other oxides which are used for
special purpose substrates are beryllia (for high thermal
conductivity), zirconia, magnesia, thoria and titania (for
high melting point), single crystal sapphire and spinel (for
uniformity of properties) and various titanates (for high
dielectric constant).

For thin film substrates the major requirement is a
high quality surface finish and, since no high temperature
processing is usually involved, glass substrates have been
extensively used.   Glasses have low thermal conductivities
and dielectric constants compared to ceramics (Table II)
hence glass ceramics, glazed ceramics and exceptionally
smooth unglazed ceramics are sometimes used (Macavoy and
Halaby, 1964; Brown, 1970).   The latter are becoming
increasingly important as substrates for microwave hybrid
circuits where materials of low dielectric loss and high
dielectric constant are required (Sobol, 1970;   Bosnell,
1971).

B.   FABRICATION TECHNIQUES

To obtain the high standard of surface finish necessary for thin film substrates it is preferable if the glass
is produced by a hot forming process, such as drawing.
Unfortunately, not all glass melts have the necessary properties to allow sheet drawing and these have to be polished.

Ceramic substrates are in a polycrystalline form and
have a wide purity and density range, depending on the
manner of fabrication.   The two most common methods are dry
pressing and band or slip casting.   In dry pressing, the
alumina and any additives (calcia, magnesia, silica, etc.)
are first mixed in a ball-mill with a binder and water to
form a slurry.   The slurry is dried to form discrete
granules of uniform size and composition which are cold
pressed into substrate dimensions.   Finally, the 'green'
substrates are sintered by firing in kilns.

In the second method, the constituents are mixed into
a slurry and a liquid or organic vehicle is added to aid
pouring.   The liquid slip is poured onto a long clean glass
plate or a continuous belt which passes under a blade to
control the film thickness.   The ceramic layer is dried
until it approximates to the green state, after which it is
cut to size and high temperature fired.   As-fired finishes
of 4-5 micro-inches are obtained for the surface in contact
with the belt or plate, and the grain size and porosity is
more uniform than for the dry pressing process.

Special quality very high density substrates can be
formed by hot pressing.   The advantage of hot pressing is
that no binders or grain growth inhibitors are required and
no impurities are deliberately added.   The substrate is of
high density and practically pore free.   It also has an

ultra-fine microstructure since the temperature used is below
the firing temperature of the other techniques and grain
growth is limited.   The surface finish is as good as that
of the forming dies.   The disadvantages of hot pressing are
size limitations and expense.

C.   SURFACE TOPOGRAPHY

Two important aspects of the surface topography which
can affect the quality of deposited films are flatness and
microfinish.   Both thick and thin film techniques require
the substrate to be flat otherwise the mechanical masks
(thin films) or screens (thick films) will not make uniform
contact and the circuit elements will be poorly defined.
The flatness achieved on drawn or float glass will depend on
the thickness and the processing temperature.   Because of
the high drawing temperature 7059 glass has a slightly wavy
surface, though in general this does not create a flatness
problem.   The only method of obtaining thin, flat and
smooth glass sheet is by polishing, which is expensive.
Similarly, thin ceramic samples usually distort during firing
and also require polishing.

The surface finish requirements for thick films are
not stringent and are mainly determined by adhesion considerations.   The surface must not be too smooth, i.e. glazed,
or the film will not adhere.   On the other hand, too rough
a surface is not good for the uniformity of film thickness.
Generally, a surface finish at 25 micro-inch C.L.A. (Centre
Line Average) is satisfactory, which corresponds to a 75-100
micro-inch peak to valley variation.

The surface microfinish for thin film circuits is much
more critical.   A single imperfection under a capacitor may

cause premature breakdown of the whole device.   Ideally,
the surface should be atomically smooth with no sharp dis-
continuities.   Drawn soda-lime and 7059 glasses have almost
perfect surfaces with only an occasional imperfection.   To
remove the detrimental influence of these defects a
dielectric film, several times thicker than the defect, can
be deposited over the surface of the drawn glass.

With the exception of those substrates produced by the
hot pressing technique, the surface of ceramic materials is
much too rough for thin film circuits and must be polished
or glazed.   Microfinishes with a C.L.A. of approximately 2
micro-inches can be obtained by careful polishing techniques,
but great care must be exercised to prevent the 'pull-out'
of individual grains which can be especially detrimental to
a thin film circuit.   The problem of pull-out can be
reduced by using substrates with extra-small grains.   These
are more difficult to remove and if pull-out does occur, the
hole size is smaller and the effect is less disastrous.
Also, higher purity alumina is easier to polish since it
contains few additives which if segregated can produce
differential removal rates.

As a result of the polishing difficulties experienced,
glazed alumina and beryllia substrates have been developed
as a compromise between the superior surface finish of glass
and the greater thermal conductivity of ceramics.   A disadvantage of the technique is that the glaze composition has
to be compatible with the ceramic and the thin film, for
example, a mismatch in the linear thermal expansion coefficients may result in bowing of the substrate and crazing
of the glaze.   Also, to prevent crystallisation at the
glaze-ceramic interface the glaze may have to contain lead

and alkalis to make low temperature application possible.
These additives may have a detrimental effect on the film
devices.

One application where the surface finish of the
ceramic substrates is particularly important is microwave
hybrid circuits.   Schilling (1968) has observed that the
surface finish has a great effect on the circuit losses of
plated-up thin film microwave circuits.   His measurements
were made at frequencies of 3, 6, and 9 GHz and are supported by the results of Bosnell (1971) at 10 GHz.   Bosnell
has also investigated the possibility of using thick film
techniques.   Here a compromise has to be made between
adhesion and a high quality surface;   the results are
significantly worse than for the thin film structure, particularly at frequencies above 7 GHz.

D.   CHEMICAL AND STRUCTURAL HOMOGENEITY

Standard glass microscope slides have been used as
thin film substrates because they have smooth fire-polished
surfaces and are cheap.   A common constituent of these
glasses is sodium oxide which under high field conditions at
elevated temperatures is highly mobile.   Such conditions
can be obtained near resistors with the result that the
alkali ions migrate to the negative terminal and interact
with the thin film structure, causing deteriation and breakdown (Siddall and Probyn, 1961).

A high alkali-metal content can also lead to a high
surface conductivity in humid atmospheres and, because of
long term leaching effects, to unstable surface conditions.
The problems associated with alkali content can be overcome
by coating the substrate with a protective thin film of

silicon monoxide before depositing the circuit.   Special
alkali free alumino-silicate glasses, Table II, have now
been developed as substrates.

It has been observed that the effects of alkali content in alumina glazes is less pronounced than for glasses.
This stability has been explained by the existence of a
lower surface temperature due to the increase in thermal
conductivity of ceramic materials, and by the high resistivity of the glaze as a result of its lead content (Leiser,
1963).

Alumina substrates consist of a crystalline phase,
mainly α-Al_2O_3, and an amorphous phase.   The amorphous
phase is generally dispersed intergranularly (Tong and
Williams, 1970) and consists of the fluxing additives,
calcia, silica, magnesia, etc.   The distribution in the
surface of the ceramic may be different from the bulk,
particularly in the case of low alumina/high additive
material.   Since most thick film inks are designed so that
the glass frit in the ink bonds to the amorphous phase the
adhesion decreases as the alumina purity increases.

Random variations in the composition and/or density of
ceramic substrates can result in thermal expansion differences which cause cracking due to localised stresses.   This
is particularly important when the substrate is subject to
non-uniform heating during certain bonding techniques.

Variations in the substrates chemical composition may
also cause losses in microwave circuits.   Bosnell has
established that the presence of alkalis, particularly Na_2O,
can contribute to the dielectric losses.   However, his
measurements of the dielectric losses obtained for a range
of substrates failed to reveal any systematic correlation

with chemical composition.   There is, however, a reasonable
correlation between electrical properties and density.
Losses are also reduced by having a lower mean grain size.
The conclusion is therefore made that the alumina powder
particle size distribution and sintering temperature are
more critical than the added impurities, provided these are
not excessive.

The purer the substrate the more uniform and stable
are its electrical properties.   If 95% alumina samples are
heated to 900°C on a belt furnace, the glassy phase may
leach out and the dielectric constant and losses are significantly changed.   Therefore the substrate requirements for a
microwave hybrid circuit are high purity, low porosity,
small grain size and smooth microfinish, in addition to a
high dielectric constant.

E.   THERMAL CONDUCTIVITY

The heat dissipated in a thin film circuit can be
removed by radiation, convection, or conduction.   Radiation
is only effective at high temperatures, while forced convection requires additional equipment;   in most cases the heat
is removed by conduction.   Since the dimensions of any
interconnections are small, the heat must be conducted away
through the substrate to a heat sink.   The thermal conductivity of glass is rather low and for this reason attempts
have been made to replace it with glass-ceramics and glazed
or hot pressed ceramics, having a much higher conductivity,
Table II.

Glass ceramics, e.g. Pyroceram, are ceramic materials
made by the controlled recrystallisation of glass.   They
have thermal conductivities about double that of glass and

can be used at higher temperatures.   The glazes that are
used to obtain smooth surfaces on some ceramics are very
thin, <0.003", and therefore do not act as a thermal barrier
except for high power circuits on an otherwise high thermal
conductivity substrate, e.g. beryllia.

The highest possible conductivity is only realised
with the high purity ceramics.   Hessinger (1963) has shown
that the thermal conductivity of beryllium oxide is
drastically reduced by minor additives through the formation
of intergranular heat barriers.   This effect is much more
significant than the decrease observed due to voids in less
dense material.   The voids are not present in all the grain
boundaries and therefore are not an effective barrier to
heat flow.

IV.   CONCLUSIONS

The substrate should provide only mechanical support
and isolation, and not interact with the deposited film
except to provide the necessary adhesion.   This condition
is only approached for the very simplest of circuits and the
degree of interaction increases with the sophistication of
the circuit requirements.   The increase in circuit complexity and the development of microwave hybrid circuits has
placed increasingly exacting requirements on the substrates
used in thin and thick film technologies.

The high deposition temperature and the reactive
nature of the semiconductors severely limits the number of
suitable single crystal substrate materials.   Only sapphire
and spinel have been extensively investigated since alternatives are not readily available.   Because of the additional
limitations imposed by the use of epitaxial techniques, more

perfect single crystal films may ultimately result from the
development of methods using amorphous substrates.

REFERENCES

Allison, J.E., Dumin, D.J., Heiman, F.P., Mueller, C.W. and
Robinson, P.H. (1969) Proc. I.E.E.E. 57, 1490.

Ang, C.Y. and Manasevit, H.M. (1965) Solid-St. Electron. 8,
994.

Arlett, R.H. and Robbins, M. (1967) J. Am. Ceram. Soc. 50,
273.

Bicknell, R.W., Charig, J.M., Joyce, B.A. and Stirland, D.J.
(1964) Phil. Mag. 9, 965.

Bicknell, R.W., Joyce, B.A., Neave, J.H. and Smith, G.V.
(1966) Phil. Mag. 14, 31.

Bosnell, J.R. (1971) Microelectronics (in press).

Braunstein, M. (1968) Proc. Am. Vacuum Soc.

Brown, R. (1970) "Handbook of Thin Film Technology"
(L.I. Maissel and R. Glang, eds.) McGraw-Hill, p.6.1.

Campbell, D.S. (1966) "The Use of Thin Films in Physical
Investigations" (J.C. Anderson, ed.) Academic Press,
London, p.11.

Chu, T.L., Francombe, M.H., Gruber, G.A., Oberly, J.J. and
Tallman, R.L. (1965) Westinghouse Res. Lab. Rep. No.AFCRL-
65-574, AD619992.

Cockayne, B. and Chesswas, M. (1967) J. Mat. Sci. 2, 498.

Cullen, G.W. (1971) J. Cryst. Growth (in press).

Cullen, G.W., Gottlieb, G.E., Wang, C.C. and Zaininger, K.H.
(1969) J. Electrochem. Soc. 116, 1444.

Doo, V.Y. (1964) J. Electrochem. Soc. 111, 1196.

Dumin, D.J. (1965) J. Appl. Phys. 36, 2700.

Dumin, D.J. (1967a) J. Appl. Phys. 38, 1909.

Dumin, D.J. (1967b) J. Electrochem. Soc. 114, 749.

Dumin, D.J. and Robinson, P.H. (1968) J. Cryst. Growth 3, 214.

Filby, J.D. (1966) J. Electrochem. Soc. 113, 1085.

Filby, J.D. and Nielsen, S. (1965) J. Electrochem. Soc. 112, 957.

Filby, J.D. and Nielsen, S. (1966) J. Electrochem. Soc. 113, 1091.

Filby, J.D. and Nielsen, S. (1967) Brit. J. Appl. Phys. 18, 1357.

Grabmaier, J.G. and Watson, B.Chr. (1968) J. Am. Ceram. Soc. 51, 355.

Haidinger, W., Courvoisier, J.C. and Kohler, W.A. (1966) Le Vide, No. spécial A.V.I. Sem. 59.

Hart, P.B., Etter, P.J., Jervis, B.W. and Flanders, J.M. (1967) Brit. J. Appl. Phys. 18, 1389.

Heiman, F.P. (1967) Appl. Phys. Letters 11, 132.

Heiman, F.P. and Robinson, P.H. (1968) Solid State Electronics 11, 411.

Hessinger, P.S. (1963) Electronics Oct.18, 75.

Holmes, P.J. and Corkhill, J.R. (1969) Electronic Components 1171.

Joyce, B.A., Bennett, R.J., Bicknell, R.W. and Etter, P.J. (1965) Trans. Metall. Soc. A.I.M.E. 233, 556.

Kelemen, D.G. (1970) Met. Trans. 1, 667.

Kingery, D.G. (1960) "Introduction to Ceramics" John Wiley and Sons Inc., New York, p.264.

Larssen, P.A. (1966) Acta Cryst. 20, 599.

Leiser, C.F. (1963) Glass Ind. 44, 509.

Macavoy, T.C. and Halaby, S.A. (1964) I.E.E.E. Trans. on Component Parts CP-11 No.1, 15.

Manasevit, H.M. (1968) Appl. Phys. Letters 12, 156.

Manasevit, H.M. and Forbes, D.H. (1966) J. Appl. Phys. 37, 734.

Manasevit, H.M. and Morritz, F.L. (1967) J. Electrochem. Soc. 114, 204.

Manasevit, H.M. and Simpson, W.I. (1969) J. Electrochem. Soc. 116, 1725.

Manasevit, H.M. and Thorsen, H.M. (1970) Metal Trans. 1, 623.

Manasevit, H.M., Miller, A., Morritz, F.L. and Nolder, R. (1965) Trans. Metall. Soc. A.I.M.E. 233, 540.

Manasevit, H.M., Forbes, D.H. and Cadoff, I.B. (1966) Trans. Metall. Soc. A.I.M.E. 236, 275.

Manasevit, H.M., Nolder, R.L. and Moudy, L.A. (1968) Trans. Metall. Soc. A.I.M.E. 242, 465.

Mercier, J. (1970) J. Electrochem. Soc. 117, 666.

Naber, C.T. and O'Neal, J.E. (1968) Trans. Metall. Soc. A.I.M.E. 242, 470.

Nielsen, S. (1960) Trans. 7th A.V.S. Symp. 293.

Nolder, R.L. and Cadoff, I.B. (1965) Trans. Metall. Soc. A.I.M.E. 233, 549.

Paul, W. and Pearson, G.L. (1955) Phys. Rev. 98, 1755.

Rai-Choudhury, P. (1969) J. Electrochem. Soc. 116, 1745.

Reynolds, F.H. and Elliott, A.B.M. (1967) Solid State Electronics 10, 1093.

Robinson, P.H. and Dumin, D.J. (1968) J. Electrochem. Soc. 115, 75.

Robinson, P.H. and Mueller, C.W. (1966) Trans. Metall. Soc. A.I.M.E. 236, 268.

Scheuplein, R. and Gibbs, P. (1960) J. Amer. Ceram. Soc. 43, 458.

Schilling, W. (1968) Microwaves 7, 52.

Schlötterer, H. and Zaminer, Ch. (1966) Phys. Stat. Sol.
15, 399.

Seiter, H. and Zaminer, Ch. (1965) Z. Angew. Phys. 20, 158.

Siddall, G. and Probyn, B.A. (1961) Brit. J. Appl. Phys.
12, 668.

Sobol, H. (1970) Solid State Technology 13, 49.

Szekely, G.S. (1969) "Nepcon 69 Proc." 605.

Tong, S.S.C. and Williams, J.P. (1970) J. Amer. Ceram. Soc.
53, 58.

Tramposch, R.F. (1966) Appl. Phys. Letters 9, 83.

Tramposch, R.F. (1969) J. Electrochem. Soc. 116, 654.

Wang, C.C. (1969) J. Appl. Phys. 40, 3433.

Wang, C.C. and McFarlane III, S.H. (1968) J. Cryst. Growth
3, 485.

Wang, C.C., Gottlieb, G.E., Cullen, G.W., McFarlane III, S.H.
and Zaininger, K.H. (1969) Trans. Metall. Soc. A.I.M.E.
245, 441.

Zaininger, K.H. and Wang, C.C. (1970) Solid-State
Electronics 13, 943.

OXIDES FOR REFRACTORY AND ENGINEERING APPLICATIONS

D.I. MATKIN

A.E.R.E., Harwell,
Didcot, Berkshire

I.   INTRODUCTION

II. THE INFLUENCE OF COMPOSITION AND MICROSTRUCTURE
     ON THE PHYSICAL PROPERTIES OF OXIDE CERAMICS
     A.   THERMAL PROPERTIES
     B.   CHEMICAL STABILITY
     C.   MECHANICAL PROPERTIES

III. THE INFLUENCE OF THE FABRICATION ROUTE ON THE
     MICROSTRUCTURE OF OXIDE CERAMICS

IV. REFRACTORY AND ENGINEERING APPLICATIONS OF
     OXIDE CERAMICS
     A.   HIGH TEMPERATURE APPLICATIONS
          EXPLOITING THE CHEMICAL STABILITY
          OF OXIDE CERAMICS
     B.   HIGH TEMPERATURE APPLICATIONS
          EXPLOITING THE MECHANICAL PROPERTIES
          OF OXIDE CERAMICS
     C.   ROOM TEMPERATURE APPLICATIONS OF
          OXIDE CERAMICS

V.   COMMENT

I.   INTRODUCTION

Oxide ceramics are compounds of oxygen containing one or more metal elements, which have sufficient stability to withstand elevated temperatures.   Since most oxides have a high degree of ionic bonding, they exhibit properties associated with ionic crystals such as optical transparency, high electrical resistivity, and low thermal conductivity. They are also characterised by refractoriness, chemical inertness, hardness, and a brittle mode of fracture.   The traditional uses of oxide ceramics have particularly exploited refractoriness and chemical stability, but other combinations of these key properties offer a class of engineering materials capable of fulfilling many modern applications.

Traditional oxide ceramics were based on clays and silicates, whilst modern technical oxide ceramics are generally pure single oxides, e.g. Al_2O_3, MgO, ZrO_2, BeO or pure mixed oxides, e.g. spinel ($MgO.Al_2O_3$), mullite ($3Al_2O_3$. $2SiO_2$), zircon ($ZrO_2.SiO_2$).   This paper discusses the properties of such polycrystalline oxides, the influence of the microstructure on these properties, and the influence of the chosen fabrication process on the microstructure. Refractory and engineering applications of oxide ceramics are also reviewed.

II.   THE INFLUENCE OF COMPOSITION AND MICROSTRUCTURE
       ON THE PHYSICAL PROPERTIES OF OXIDE CERAMICS

The most important parameters which control the micro-structure of polycrystalline ceramics are chemical composition (impurities), porosity, and grain structure.

Fig.1 shows a schematic representation of microstructure of three different types of oxide ceramics.   Microstructure A represents a very pure single oxide (99.5% Al_2O_3) containing a small amount of porosity (1%) at the grain boundaries. Microstructure B represents a pure single oxide (99% Al_2O_3) with 1% of impurity precipitates, and rather more porosity (5-10%) than type A.   Microstructure C represents a debased oxide (95% Al_2O_3) in which the individual grains are bonded by a matrix of a glassy phase formed by the 5% impurities.

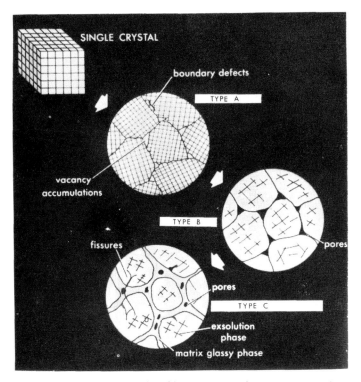

Fig.1. Schematic diagram of three types of microstructure typical of oxide ceramics.

A.   THERMAL PROPERTIES

The outstanding property of most oxide ceramics is a
high melting point, but the ability to withstand high
operating temperatures in a variety of atmospheres is
determined by both melting point and chemical stability.
The high melting point of oxides is associated with the
degree of covalency of the atomic bonding.   (ThO$_2$ is the
most refractory oxide known with a melting point of 3300°C.)

The thermal conductivity of most oxide ceramics is
somewhat lower than most metals and ceramic carbides and
decreases with increasing temperature.   The thermal conductivity of a range of polycrystalline oxides (corrected to
theoretical density) at various temperatures are shown in
Fig.2.

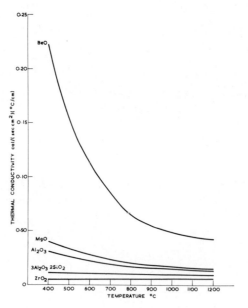

Fig.2.   Thermal conductivity of
oxide ceramics vs temperature.

The fall in thermal conductivity with temperature, between $100^{\circ}C$ and $1000^{\circ}C$, is fairly dramatic for BeO, MgO and Al_2O_3; for ZrO_2 (stabilised) there is a very slight increase. The high thermal conductivity of BeO at $100^{\circ}C$ is equivalent to that of aluminium metal but small amounts of impurities have a marked effect, e.g. 4% of impurities reduce the thermal conductivity at $100^{\circ}C$ by nearly 50% whereas 5% of porosity in pure BeO only reduces the thermal conductivity at $100^{\circ}C$ by 5%.

The coefficient of thermal expansion of most oxide ceramics is somewhat lower than metals, ranging from 14×10^{-6} $^{\circ}C^{-1}$ for SrO to 5×10^{-6} $^{\circ}C^{-1}$ for $Al_2O_3.TiO_2$, and even negative values occur for certain lithium compounds such as β-spodumene ($Li_2O.Al_2O_3.4SiO_2$). The coefficient of thermal expansion of oxide ceramics is important since the ability to withstand thermal shock is primarily determined by the ratio of fracture strength to expansion coefficient. Under certain conditions thermal conductivity is the more important parameter, since a high conductivity reduces the temperature gradient within the component.

B.   CHEMICAL STABILITY

The refractory oxides include some of the most stable compounds known, with respect to both a gaseous environment and resistance to attack by molten metals, etc. Stability in a specific environment may be limited by a tendency to oxidise, reduce, volatilise or hydrate at high temperature. In oxygen or air, the oxides MnO, UO_2 and U_2O_3 oxidise to higher oxides, but in pure hydrogen nearly all oxides are stable with the exception of NiO which is reduced to Ni metal. The oxides MgO, CaO, SrO, BaO, MnO and NiO have

relatively high dissociation pressures and can be readily
volatilised in vacuo above 1700°C; MgO is known to dis-
sociate rapidly above 1600°C under a vacuum of 10^{-4} torr.
BeO has a marked volatility in the presence of water vapour
at temperatures as low as 1500°C due to the formation of a
stable gaseous $Be(OH)_2$ molecule.

Although Al_2O_3 has a lower melting point than many
oxides (2050°C), it is one of the most stable under a wide
variety of atmospheres.   These properties combined with
relative cheapness and availability make it a widely used
oxide ceramic.

C.   MECHANICAL PROPERTIES

Mechanically, oxides have three outstanding attributes;
high hardness (hence good wear resistance), high stiffness,
and high strength (at room and elevated temperatures).
Disadvantages are, however, brittle behaviour with poor
resistance to thermal and mechanical shock.

The wear resistance of polycrystalline oxide ceramics
is not well understood.   Mechanisms include the wear of
individual oxide grains and the extraction of whole grains.
The influence of microstructure on wear resistance has
received very little attention.   The stiffness, or elastic
modulus, of polycrystalline oxide ceramics is generally high,
e.g. alumina, which has a maximum value of 400 GN/m^2.   The
elastic modulus is influenced by microstructure, porosity
and impurities being particularly important.

The strength of polycrystalline ceramics has been
reviewed by Davidge and Evans (1970), and the influence of
microstructure on strength is understood in a qualitative
manner.   If oxides are assumed to be truly brittle, their

strength can be determined by the modified Griffith equation

$$\sigma_f = k\sqrt{(E\gamma_i/c)}$$

where σ_f = fracture strength, E = elastic modulus, γ_i = effective surface energy, c = flaw size, and k = geometrical constant.   Flaws can arise either from intrinsic flaws arising during fabrication, at the surface by damage, or from interactions between dislocations if the material is plastically deforming.   For most oxides at room temperature, the finer the microstructure, the smaller the size of the flaws, and the higher the tensile or bend strength.   Values of bend strength of 70 MN/m^2 have been attained for high density pure polycrystalline Al_2O_3 with an average grain size of 1-2µm (type A, microstructure).   Clearly the high temperature mechanical strength of a debased oxide (type C, microstructure) will be determined by the softening point of the glassy matrix, and in debased Al_2O_3, for instance, the strength decreases rapidly above 1000°C.

Although oxide ceramics can have high strengths they are often brittle and lack toughness, exhibiting a catastrophic mode of fracture.   The mode of fracture is very insensitive to changes in microstructure, and it is necessary to incorporate either fibres or whiskers in a ceramic composite to achieve an increase in toughness (Matkin, 1970).

III.   THE INFLUENCE OF THE FABRICATION ROUTE ON THE
       MICROSTRUCTURE OF OXIDE CERAMICS

The microstructure of any polycrystalline oxide ceramic is determined by the total fabrication process from the preparation of the initial powder to the final post-

sintering operations (Matkin, 1970). The aim of any
fabrication route is threefold: to obtain the required
microstructure, to obtain the required shape with given
dimensional tolerances, and to produce economically. The
fabrication routes for oxide ceramics can be divided into
three main groups, (1) cold forming + high temperature
sintering, (2) hot forming, (3) melt processes (including
ceramic coatings).

The traditional route is cold forming oxide powders
followed by a high temperature firing process to achieve
strength, during which shrinkage occurs. Fig.3 summarises
the steps in this route, and indicates the four types of
cold forming used to achieve the required shapes.

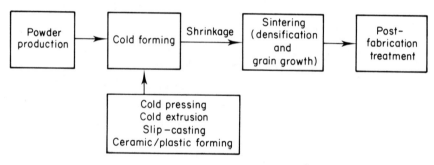

*Fig.3. Stages in conventional
fabrication of ceramics.*

Cold pressing of powders includes straight pressing
and cold isostatic pressing in rubber bags surrounded by a
pressurised fluid. In the last 10 years the technology of
isostatic pressing has been transferred from the laboratory
to the ceramics industry, and large blocks of alumina can
now be readily prepared. A typical size is 35 x 45 x 250cm.

Cold extrusion of oxide compositions with claylike consist-
ency is an ideal process for production of tubes and bars,
while slip casting of aqueous suspensions of oxide particles
is more suited to the fabrication of thin walled complex
shaped articles.   Various warm forming techniques employed
in the plastic industry (injection moulding, transfer
moulding, etc.) can be applied to mixtures of plastic and
oxide powder after which the plastic is burnt out, and the
oxide components fired in the normal way.   These processes
allow the fabrication of complex shaped components.

During the high temperature firing, densification
occurs by processes such as diffusion of vacancies from
pores to grain boundary sinks.   Grain growth is also
possible and may be 'normal' (the average grain size
increases gradually), or 'discontinuous' (one or two grains
grow catastrophically trapping porosity within the grains)
as illustrated in Fig.4.   The presence of a few large
grains will have a marked influence on the strength of the
components, and additive impurities have been empirically
developed to prevent discontinuous grain growth.   The
'Lucalox' process, in which a small quantity of MgO is
incorporated into Al_2O_3 during fabrication, is the best
known additive (Coble, 1961).   Fig.4a shows zero porosity,
and in this condition the material is translucent; a
similar translucency has been achieved in BeO (Denton et al.
1969), whilst fully densified Y_2O_3 and MgO are transparent
(Jorgenson and Anderson, 1967; Miles et al. 1967).   The
linear shrinkage during firing can be as high as 20%, and
hence as-fired dimensional tolerances of $\pm$ 1% are standard;
subsequent diamond grinding can be used to achieve very high
tolerances.

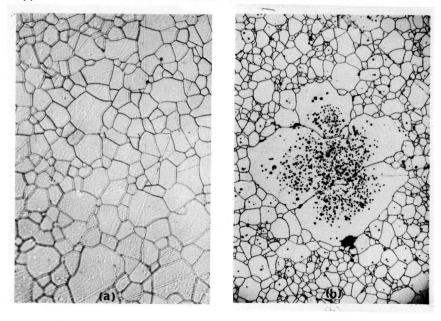

Fig.4.   Microstructure of dense Al$_2$O$_3$ illustrating
    the influence of MgO on grain growth.
a)   Uniform microstructure due to addition of MgO.
b)   Discontinuous grain growth in pure Al$_2$O$_3$.

If a microstructure is required with high density and
very fine grain size (high room temperature strength), it
may be necessary to apply pressure during the sintering of
the powder (hot-pressing).   Hot forging of dense ceramic
bodies has also been developed to allow fabrication of
complex shapes with high density fine grain microstructure
(Spriggs et al. 1969);   the resultant microstructure
perpendicular to the forging direction is different to that
parallel to the forging direction.

The group of melting processes includes fusion casting,
in which the oxides are cast in a molten condition, the
casting of special glass compositions which are subsequently

devitrified into a glass ceramic (such as β-spodumene), and
the deposition of oxide coatings by the production of molten
oxide droplets in a flame or plasma torch.

IV.   REFRACTORY AND ENGINEERING APPLICATIONS OF
      OXIDE CERAMICS

Most of the applications to be discussed exploit a
combination of two or more key physical properties.
Refractoriness of oxides is not always the most critical
property, and the applications have therefore been divided
into two groups at high temperature and one at room temperature.   Applications of electrical and magnetic oxides are
not discussed since these have been fully covered elsewhere.
Also, there will be no detailed discussion of applications
of oxides as nuclear fuels, nuclear moderators, or as structural components in nuclear reactors.   It is sufficient to
note that nuclear applications of oxides exploit their
refractoriness and resistance to nuclear irradiation.
These oxides are often referred to as nuclear refractories,
and apart from BeO and B_4C have no uses outside the field of
nuclear energy.   However, much of the present day understanding of oxide properties, and the technology of
fabrication of technical oxide ceramics has been derived
from the need for very high quality nuclear ceramics, for
example high purity, high density, fine grain size UO_2 and
BeO.

A.   HIGH TEMPERATURE APPLICATIONS EXPLOITING THE
     CHEMICAL STABILITY OF OXIDE CERAMICS

In these applications, refractoriness and resistance
to chemical attack are the important physical properties.

1. Oxide refractories

The largest single application for oxide ceramics is for refractories in the iron and steel, glass, cement, and gas industries.   The present annual consumption of such refractories is 1.5M tons, with a market value of around £12M.   Cost is still a very important factor since the average consumption of refractories is 0.04 tons of refractories per ton of steel, at a cost of $\sim$ 50p per ton of steel.

The basic application of a refractory is as a structural component operating at high temperatures, and subjected to rapid heating and cooling cycles whilst in contact with corrosive gases and liquid slags.   The U.K. steel production can be divided into three processes; open hearth furnaces, electric arc furnaces, and the basic Bessemer process.   During the last ten years the pattern of steel making has changed from approximately 90% for open hearth and zero% for oxygen converters in 1958, to 55% in open hearth and 30% in L.D. and Kaldo oxygen converters. In these changes the classical refractories, silica and fire clay, have been replaced by basic refractories (chrome-magnesite, magnesite, magnesite-chrome and dolomite, in that order of usage).   Basic refractories can operate at higher temperatures (2000°C), are more resistant to chemical attack from alkaline slags, and are able to withstand the erosive conditions experienced in L.D. and Kaldo vessels.   In addition, high alumina refractories with high operating temperatures (1800°C) have replaced silica refractories (1700°C) in open hearth and electric arc furnaces.   These high performance refractories have a number of properties in common:

(i)   Refractoriness.

(ii) Volume stability up to the operating temperature.

(iii) Good chemical resistance to the environment.

(iv) Good mechanical properties up to the operating temperature.

The ability of any refractory to meet these demands is determined by the microstructure, e.g. impurity phases, porosity, etc.   The maximum operating temperature is determined by the major solid phase, i.e. SiO_2, Al_2O_3, or MgO, but the temperature at which the first liquid is formed, which may be well below the operating temperature, is determined by impurity phases.   The presence of liquid phases within the refractory at the operating temperature greatly influences the high temperature mechanical properties.   The amount and type of porosity influences both the mechanical properties, and the ability of the refractory to withstand chemical attack via penetration of the pores.

The developments of improved refractories in recent years has resulted from the use of higher purity bricks (increased refractoriness and chemical resistance), the fabrication of direct or chemical bonding between grains within the microstructure (improved high temperature mechanical properties), and the use of tar impregnation to increase the resistance to slag attack by penetration. Advances in fabrication technology have allowed the manufacture of refractories with improved microstructure, more complex shapes, and larger components.

Other types of refractories find use in specific applications. For example, fusion cast refractories are employed in the side walls of electric arc furnaces, in glass melting tanks and regenerators;   the annual U.K.

market in fusion cast refractories is approximately £1M.
Another limited application of oxide ceramic refractories is
in the processing of non-ferrous metals, such as liquid
aluminium, where the usefulness of a refractory is limited
by the degree of wetting by the molten metal.

2.   Chemical and metallurgical applications

The most demanding of these applications is the use of
oxide crucibles for the containment of molten refractory
metals.   The general stability of oxides in contact with
metals at high temperature may be listed in the following
order of decreasing reactivity: ThO_2, BeO, ZrO_2, Al_2O_3, and
MgO.   In addition to chemical inertness, ThO_2, BeO, and
ZrO_2 have relatively low vapour pressures and can be used in
vacuum casting of molten metals.   In addition to compatibility, the degree of melting of the oxide by the liquid
metal is important in determining the usefulness of the
ceramic.   Other chemical and metallurgical applications
include furnace tubes and thermocouple sheaths, where in
addition to refractoriness and chemical stability it is
important to have high electrical resistivity and high
thermal conductivity at the operating temperatures.

Oxide ceramics are found in a wide variety of applications in the chemical processing industries.   In the
simplest examples they are used as structural components in
a high temperature chemically corrosive atmosphere.   More
specific applications are Al_2O_3 balls or pebbles as tower
packing for the high temperature heat exchanger in various
chemical processes, such as steam generation, shale oil
recovery, and the cracking of crude oils. Al_2O_3 components
with relatively high surface areas are used as supports for

catalysts in a number of chemical processes.

A more recent application, still in the development stage, is that of an Al_2O_3 honeycomb as a catalyst support for the control of air pollution (Acres, 1970).   Honeycomb materials have been developed consisting of thin corrugated Al_2O_3 sheets bonded to form a rigid honeycomb structure with a high surface to volume ratio.   This structure, combined with new techniques developed to preferentially impregnate only the surfaces of the corrugated sheets with the catalytic platinum, allows economic treatment of polluted air to be envisaged.   The applications include abatement of nitrogen oxide mixtures from nitric acid plants, elimination of organic fumes from a variety of industrial processes (paint baking ovens, plastic processing, etc.) and most important, the elimination of CO and unburnt hydrocarbons from the exhaust fumes of diesel and internal combustion engines.

3.   Automobile applications

A major application of oxide ceramics is the use of Al_2O_3 insulators in automobile sparking plugs and jet igniters;   a debased 95% alumina with a microstructure of type C (Fig.1) is normally used.   The annual U.K. market for sparking plugs is approximately £2M.   In addition to the obvious requirement of electrical insulation at high temperature, there are a number of other demanding requirements.   The insulating material must withstand temperatures up to $900°C$ with rapid temperature changes between induction and ignition strokes, and pressures of up to 900 p.s.i. with rapid pressure changes.   Good thermal conductivity is required, and the Al_2O_3 must also withstand the corrosive

gaseous environment particularly from the use of lead containing anti-knock fuels.   As cost is important, the
fabrication process has been fully automated, including the
introduction of automatic isostatic pressing machines.

A novel application for ceramics in the automobile,
still at an early stage of development, is the use of a
glass-ceramic heat exchanger in gas turbine engines (Penny,
1966).   The requirements for a rotatory heat exchanger
material for such an engine are: (a) potentially low cost
when formed into thin walled matrix structures, (b) high
temperature strength, (c) good dimensional stability at
temperature, (d) low density, (e) low thermal conductivity,
(f) high specific heat, (g) good oxidation resistance.

Several oxide ceramics could meet most of these requirements, but a material is required with good dimensional
stability over a wide temperature range.   Dimension
stability is required to allow high pressure rubbing seals
on the rim of the heat exchanger disc, and since the disc is
∿ 42cms diameter it is necessary to use a material with an
extremely low coefficient of thermal expansion.   Glass-
ceramics can be fabricated with compositions (lithium
aluminium silicates) that have virtually zero thermal expansion, and suitable heat exchanger structures have been
developed and tested in working engines.

4. Miscellaneous applications

It was shown earlier that transparent or translucent
polycrystalline oxides can be obtained when porosity is
close to zero.   The most important high temperature application of such oxides is as tube material for high efficiency
light sources.   The established sodium vapour lamp consists

of a silica tube filled with sodium vapour, operated at ∿
1000°C.   The use of alumina 'Lucalox' tubes with an optical
transmission of 90% in the visible region, combined with
refractoriness and resistance to chemical attack, allows the
sodium vapour to be operated at higher pressures and higher
temperature (∿ 1300°C).   Trials of these new sodium lamps
(claimed to produce 100 lumens/watt compared with 22 lumens/
watt for incandescent lamps) are now being carried out.
Single crystal sapphire tubes 'Saphikon' are now available
in sizes up to 1 inch diameter, and are being considered as
an alternative to translucent polycrystalline Al_2O_3 for
sodium vapour lamps since their optical transmission is
approximately 98.5%

B.   HIGH TEMPERATURE APPLICATIONS EXPLOITING THE
      MECHANICAL PROPERTIES OF OXIDE CERAMICS

      In these applications the key properties are refrac-
toriness and mechanical strength.

      The problems that arise during the introduction of a
new technology are illustrated by oxide ceramic cutting
tools (King and Wheildon, 1966).   In the early 1950's the
wide application of oxide ceramics as tool tips for machining
metals was predicted, and several ceramic tools were introduced with outstanding claims.   Today the metal cutting
market is equally divided between high-speed steel tools and
cemented tungsten carbide tools, with ceramic tools providing
less than 1% of the market.   However, the market for oxide
ceramic cutting tools is expanding.   The first oxide ceramic
tools were debased alumina (90% Al_2O_3) with a type C microstructure (see Fig.1), resulting in a room temperature
modulus of rupture of 35 MN/m^2 and poor high temperature

strength.   Advances in the fabrication of high density, fine grain size, pure alumina (99.5% Al_2O_3) have resulted in a material of type A (see Fig.1) with a rupture modulus of 70 MN/m^2 up to 1000°C.   There have been parallel improvements in wear resistance, compressive strength, elastic modulus, resistance to oxidation, and chemical reaction with metals.   Modern Al_2O_3 cutting tools are available as disposable tool tips in a variety of shapes and sizes, and can offer savings as a result of increased wear resistance and hence longer life, higher removal rates, and better as-machined surface finishes which eliminate expensive grinding. The wide use of oxide cutting tools is restricted by the tendency to catastrophic failure if subjected to impact during interrupted cutting, or contact between the tool tip and a slag inclusion in the metal components.   Nevertheless, the commercially available oxide tools are finding increased application.

An associated application for Al_2O_3 is the use of fused Al_2O_3 grains or granules as an abrasive material in grinding wheels, grinding papers and lapping and polishing powders. Al_2O_3 raw material is fused in a carbon electrode furnace, and the fused lumps are crushed, ball-milled and sieved before being incorporated into grinding wheels with either a metal, plastic, or clay-type matrix.

The high temperature hardness and wear resistance of Al_2O_3 is also exploited in ceramic wire drawing dies and cones.   Typical cones suitable for drawing copper wire are shown in Fig.5;   a further advantage is that dense Al_2O_3 can be machined to give a surface finish of around 10μ"C.L.A.

Al_2O_3, SiO_2, BeO and glass-ceramics are also used as radomes for rockets and missile noses where the material must have given dielectric properties, good mechanical resistance to corrosion by impact of rain drops, and refractoriness to withstand high temperatures on re-entry.   The dimensions of radomes are very critical, and the established fabrication routes are slip casting or flame-spraying followed by sintering, or, in the case of glass-ceramics, by the casting of molten glass and controlled de-vitrification.

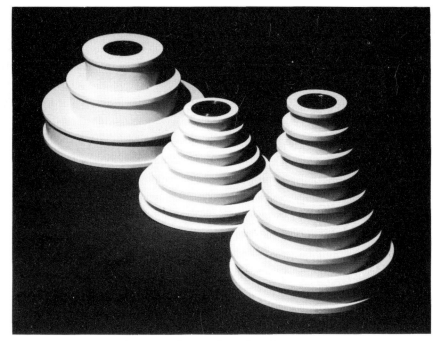

Fig.5. Al_2O_3 wire drawing cones
(Courtesy of Smiths Industries Limited).

C.   ROOM TEMPERATURE APPLICATIONS OF OXIDE CERAMICS

These applications involve temperatures up to $100^{\circ}C$ and the major properties exploited are hardness and high wear resistance.

A similar application to the high temperature wire drawing cones is the use of oxide ceramics for textile thread guides. The increased speed of modern textile machinery, and the increased total throughput of abrasive natural and man-made fibres has led to the manufacture of a wide range of Al_2O_3 thread guides (Fig.6).

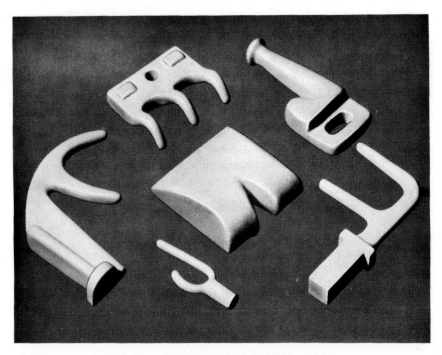

Fig.6. Al_2O_3 textile thread guides
(Courtesy of Smiths Industries Limited).

These complex shapes are generally fabricated by injection moulding and subsequent finishing operations achieve surface finishes of $\sim 10\mu''$C.L.A. In certain applications, electrically conducting TiO_2 thread guides are used to prevent build-up of static electricity.

There is an enormous list of applications of oxide ceramic components (mainly Al_2O_3) under the general heading of ceramic bearings, ranging from oxide ceramic gyro bearings, pump seals in washing machines, automobile water pumps, and in pumps for corrosive chemicals, through to sapphire watch bearings. For many bearing applications it is possible to coat a metal component with a thin ceramic coating. Provided the ceramic composition is intrinsically hard and has good bonding to the metal substrate, satis- factory bearing surfaces can be produced either by plasma/ flame spraying or by a vapour deposition technique (Hayden, 1969).

A further range of applications of Al_2O_3 arise from the high resistance to abrasion by hard particles and include shot blasting nozzles, extrusion dies used to extrude pipe from highly abrasive flint-loaded clay, and cyclone centrifugal filtration cones used in separating slurries of highly abrasive materials. An associated application is the use of Al_2O_3 ball-mills and balls in the production of fine powders not only in ceramic fabrication, but also in the production of paint pigments, talc, etc.

A novel application of pure Al_2O_3 is for deep sub- mergent buoys used in suspending oceanographic equipment at depths of up to 10,000m. Hollow spheres up to 25cm diameter have been fabricated; the high compressive strength of Al_2O_3 allows the use of very thin walls for maximum buoyancy. Such spheres give long service life due to their chemical stability in salt-water.

Oxide ceramics are also used as either personnel armour or transparent window armour where the important properties include low density, good ballistic properties (determined by elastic properties and hardness) and the ability to be fabricated into complex shaped sections. The personnel armour unit is a composite of an Al_2O_3 plate backed by a glass-fibre reinforced resin. The success of the composite depends on the reflection of the shock wave at the backing, which leads to ejection of the projectile and often a cap of ceramic, away from the front face of the ceramic plate. Transparent armour for use in helicopter windows must have good ballistic properties, and obviously must be transparent.

V.   COMMENT

The use of oxide ceramics as refractory and engineering materials exploits various combinations of the key physical properties. Further applications of these outstanding properties depends upon engineers accepting a philosophy of designing with brittle materials. So far the term ceramic has been synonymous with oxides, but more recently several new attractive non-oxide materials such as Si_3N_4, and SiC, have been developed. These are likely to compete with oxides in future applications.

REFERENCES

Acres, G.J.K. (1970) Platinum Metals Rev. 14, 2.

Anon (1968) Engineering, 13th December, 1968, 898.

Coble, R.L. (1961) J. Appl. Phys., 32, 787 and 793.

Davidge, R.W. and Evans, A.G. (1970) Mat. Science and Eng. 6, 281.

Denton, I.E., Matkin, D.I. and Hill, N.A. (1969) Proc. Brit. Ceram. Soc. 12, 33.

Hayden, C.W. (1969) Ceramic Age, January 1969, 40.

Jorgenson, P.J. and Anderson, R.C. (1967) J. Am. Ceram.
  Soc. 50, 553.

King, A.G. and Wheildon, W.M. (1966) "Ceramics for Machining
  Processes", Academic Press, New York.

Matkin, D.I. (1970) Science of Ceramics 5, 441.

Miles, G.D., Sambell, R.A.J., Rutherford, J. and
  Stephenson, G.W. (1967) Trans. Brit. Ceram. Soc. 66, 319.

Niesse, J.E. (1970) S.A.M.P.E. 1 (4), 17.

Penny, N. (1966) Society of Automotive Engineers, Meeting
  in Detroit, U.S.A., June 1966.

Spriggs, R.M., Runk, R.B. and Atterrass, L. (1969) Proc.
  Brit. Ceram. Soc. 12, 65.

OXIDES IN COMPOSITES

B. HARRIS

University of Sussex,
Brighton, Sussex

I.   INTRODUCTION

II.   OXIDES AS STRONG SOLIDS

III. MANUFACTURE AND PROPERTIES OF OXIDES IN HIGH
        STRENGTH FORM
        A. WHISKERS
        B. POLYCRYSTALLINE FIBRES
        C. GLASS AND SILICA FIBRES

IV.   STRONG BUT BRITTLE SOLIDS

V.   PARTICULATE COMPOSITES

VI.   FIBRE-REINFORCED COMPOSITES
        A. MANUFACTURE OF FIBRE-REINFORCED COMPOSITES
        B. PROPERTIES OF REINFORCED METALS
        C. PROPERTIES OF REINFORCED PLASTICS
        D. TOUGHNESS OF FIBRE-REINFORCED COMPOSITES
        E. FIBRE REINFORCED CERAMICS

I.   INTRODUCTION

Oxides have many desirable properties, high strength,
high rigidity, thermal and chemical stability, but as far as
the engineer is concerned their use is limited by undesirable

258

attributes such as intense brittleness. However, by
combining brittle oxides with more tractable materials the
designer can make use of favourable properties without
suffering unduly from brittleness. The reinforcing oxides
may be crystalline or amorphous and in the form of powder or
fibre; the fibres can be fine and short, or thick and continuous. The matrix may be a soft metal or plastic or
another brittle material such as a glass. Alternatively,
the brittle oxide can be used as a matrix for ductile
metallic fibres or particles. Although it would appear
that the possible number of composite combinations is vast,
in practice the range is restricted because the choice of
useful oxides is rather limited. Most of this discussion
is concerned with metals and plastics reinforced with
alumina or silica in the form of particles, single crystal
'whiskers', polycrystalline fibres, or glass fibres.

II. OXIDES AS STRONG SOLIDS

Attempts have been made to calculate the theoretical
strength of a crystalline solid, or of the random network of
chain-like molecules in a glassy solid, and to determine the
factors responsible for the high strengths of strong solids.
Ideal strong solids have a high Young's modulus, high surface energy, and small separation of the atoms. These
properties are related to each other and are determined by
the type of chemical binding: they are found in solids
bound by covalent bonds or by strongly-polarised ionic bonds.
The highly-directed nature of these bonds also ensures a
high degree of torsional rigidity. Fig.1 illustrates that
the elastic moduli of many families of solids are inversely
related to the separation of the atoms or ions. Hence, in

order to obtain a high modulus, and high strength, it is
necessary to have small atoms with short bond lengths so
that a large density of bonds per unit volume is produced.
In covalent solids a valency of 3 or 4 is necessary to
ensure that a three-dimensional network is produced instead
of molecular chains or simple molecules.   In ionic solids
small ions of high charge ensure that bonds are highly-
polarised.   The elements which fulfil these requirements
are Be, B, C, N, Al, Si and O, and the strongest solids
always contain one of these elements and frequently only
these.   Oxides are therefore potentially some of the most
useful of reinforcing materials.   Further consequences are
that strong solids will also be materials of low density and
high melting point, factors of importance in materials for
minimum weight design or for high-temperature operation.
Chemical stability, or resistance to oxidation, of materials
which are already in a highly oxidised state, is a further
bonus.   Fig.2 (Olds, 1966) shows how oxides in the bulk
state compare with other materials.

        A consideration of theoretical strength alone does not
give the whole picture, for strengths of strong solids are
often lower than the theoretical values.   All solids contain defects, such as dislocations, which may be mobile and
can cause the solid to shear under low stresses.   This is
true of metals and ionic solids of low charge, but in oxides
and other ceramics the structure of dislocations is complex
and they are usually immobile, except at high temperature,
when diffusion is able to assist motion.   Besides these
packing defects, all solids contain flaws such as surface
cracks or steps, or internal defects such as pores, any of
which may be introduced by manufacturing or handling.   In

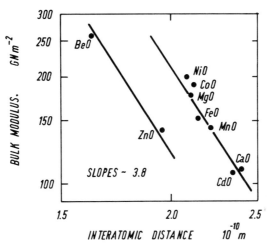

Fig.1. *The effect of atomic spacing on the elastic stiffness of some oxides.*

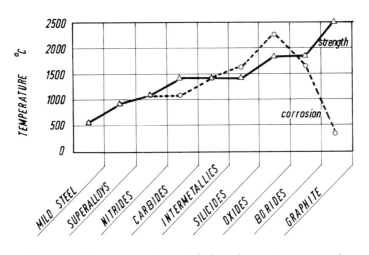

Fig.2. *The temperature limitations for several classes of material, imposed by strength and atmospheric corrosion (after Olds, 1966).*

TABLE I

Material	Physical Form	Particle or Fibre Diameter, (Microns)	Density, (10^3 Kg/m^3)	Young's Modulus, E, (GN m^{-2})	Tensile Strength σ_f (MN m^{-2})	σ_f/σ_{max}
Al_2O_3	Theoretical	-	-	460	46,000 $(= \sigma_{max.})$	-
	Sintered, bulk polycryst. body	$> 10^4$	3.85 - 3.92	350	280	1/200
	Fully-dense, bulk polycryst. body	$> 10^4$	3.98	370	700	1/60
	Polycryst. fibre	100	3.15	370	2,000	1/25
	Linde single crystal, ground rod	10^4	3.98	400	500	1/100
	Linde single crystal, flame polished rod	10^4	3.98	400	5,000	1/9
	Whisker	10	3.98	490	7,000	1/7
	Whisker	1	3.98	490	21,000	1/2
Glass & SiO_2 (non-crystalline)	Theoretical	-	-	73	16,000	-
	Bulk glass, off the shelf	$> 10^4$	2.5	70	7 - 140	1/1,000
	Ordinary quality E-glass fibre	10	2.5	70	1,500 - 2,000	1/10
	Etched E-glass	$\sim 10^3$	2.5	70	2,800	1/6
	Specially prepared E-glass fibre	5 - 50	2.5	70	3,700	1/4
	S-glass fibre	10	2.6	84	4,550	1/3
	SiO_2 fibre, tested in air	50	2.2	70	5,600	1/3
	SiO_2 fibre, tested in vacuum	50	2.2	70	7,000	1/2

strong solids these defects are responsible for failure at
relatively low stresses.   The effects of such flaws can be
minimised, so that considerable fractions of the ideal
strength can still be realised.

Pure oxide ceramics are shaped for engineering purposes by pressing graded powders.   Pressing may be carried
out cold, followed by a high-temperature sintering operation
to induce diffusion across particle boundaries and assist in
the elimination of voids.   Alternatively, hot pressing
produces a better quality, denser body at lower temperatures
and pressures than when the operations are carried out
separately.   However, the component still contains grain
boundaries, internal voids which can never be completely
eliminated by sintering, and some cracks left over from the
comminution process.   Impurity particles or films may also
remain at the grain boundaries.   Due to stress concentration,
these defects are able to extend at low applied stresses, by
breaking bonds a few at a time and, as a result, bulk oxides
are weak.in tension (Table I).   The severity of these flaws
is quite random, so one piece of material might break at
some stress, σ_a, whilst another could easily break at half
or twice that stress.   It is therefore necessary to treat
the data from strength tests by statistical methods.   This
leads to large, wasteful, safety factors and a reluctance to
use ceramic materials in engineering unless the component is
stressed mainly in compression.

III. MANUFACTURE AND PROPERTIES OF OXIDES IN HIGH
      STRENGTH FORM

One simple way to obtain oxides in a relatively strong
form is to make use of flaws to reduce the solid to powder.

Each time a piece of brittle solid breaks, the most dangerous
of the existing flaws is used up, and the remaining pieces
are consequently stronger. When the solid is reduced to a
fine powder, each particle will probably be much stronger
than the bulk solid; if the ceramic is retained in this
form by incorporating it in a metal or plastic matrix its
higher strength can be utilised.

        Large single crystals contain none of the manufacturing
defects present in sintered bodies. Table I shows that
these crystals can be very strong if the surfaces are carefully prepared, e.g. by flame-polishing, so that they are
free from cracks and steps. Large, fully dense, polycrystalline bodies of MgO and Al_2O_3 have also been made by
hot pressing with additives to aid sintering. The result
is that internal flaws can be largely eliminated, and grain
boundaries are no longer the sources of weakness as in conventional sintered bodies. But ceramics in single crystal
and fully-dense polycrystalline form are very costly and are
still susceptible to weakening by surface damage.

A.   WHISKERS

        One of the strongest forms in which materials are
found initially proved troublesome in miniaturised circuits.
Whisker growth on metal surfaces, which led to short
circuiting, consisted of fine highly perfect single crystals.
They contained no grown-in dislocations and their strengths
approached theoretical estimates for perfect solids. Such
whiskers may be a few microns thick and a tenth of a millimetre long. Much effort has been expended on studies of
the growth and properties of whiskers and on means of
incorporating them into metallic or plastic matrices. Most

strong solids have been grown in whisker form.

In early studies of Al_2O_3 whisker growth a batch
process was used in which aluminium was heated in a porcelain
boat in a stream of wet hydrogen at about 1250°C (Webb and
Forgeng, 1957).   Modern methods use dry hydrogen and a
mullite refractory boat, and the growth is considered to be
the result of a reaction between aluminium and silicon monoxide, in vapour form, which is formed by the reduction of
silica from the refractory boat.

Several other methods can be successfully used for
whisker production and the size, shape, purity and mechanical properties of the whiskers are very sensitive to the
details of the manufacturing process.   Whiskers cannot be
used in the condition in which they first grow: they must
be separated and graded into fractions within which there is
a limited distribution of the length-to-diameter, or aspect
ratio.   In this process, and during incorporation into a
suitable matrix, it is important to avoid damaging the
fibres.

The mechanical properties of aluminium oxide whiskers
are well documented, but whiskers of other oxides have
received little attention. α-Al_2O_3, or corundum, is the
form of 'sapphire' whisker studied most.   It is hexagonal
in structure, and growth occurs in four common directions
referred to as A_1, A_2, A-C and C type crystals.   At low
magnifications the whiskers appear to be smooth and straight
but closer inspection shows that the surfaces are frequently
rough.   This roughness is a source of stress concentration
and weakness, which added to the difficult task of testing
such minute fibres accounts for the scatter in reported
values of strength.   For the more common A-type whiskers

the strength can be expressed by the relation

$$\sigma_f = KA_s^{-n}$$

where A_s is the surface area of the whisker and n is a
material constant with a mean value of about 0.5.   It is
believed that the strength of A-type whiskers is governed by
surface flaws, whereas the strength of C-type whiskers obeys
a relationship of the type

$$\sigma_f = Kd^{-0.64}$$

and does not depend on the fibre length (Bayer and Cooper,
1967).   Most whiskers of about 10μ effective diameter have
strengths $\sim$ 7000 MN m^{-2}.   Strengths up to 21,000 MN m^{-2}
have been recorded for whiskers of 1μ diameter but the
scatter is considerably greater and mean values remain $\sim$
10,000 MN m^{-2}.   Early measurements of Young's modulus suggested that this was also dependent on fibre cross-section,
but recent work has shown that in testing very small
whiskers machine errors and size determination errors become
significant; the modulus does not in fact vary with size.
It is currently believed that the modulus of sapphire
whiskers is no higher than that for bulk sapphire, which is
of the order of 450 GN m^{-2}.

B.   POLYCRYSTALLINE FIBRES

    An important material design parameter is fracture
toughness.   In fibre-reinforced composites toughness has
been shown to increase with increasing fibre diameter
(Cooper and Kelly, 1967) and manufacturers have sought to
produce reinforcing fibres of oxides and other ceramics with

small diameters of the order of tenths of a millimetre.
Larger diameter fibres are likely to be more stable than
whiskers at high temperatures because whiskers often lose
their integrity when heated, especially in contact with some
of the more common matrix materials such as nickel alloys.
Single crystals and polycrystalline fibres have been made by
batch and continuous methods. One method used has been
conventional ceramic processing by extruding aqueous or
plastic-based mixes of alumina powders, which are then dried
and fired at temperatures up to 1900°C. Melt-forming
methods have also been used. These fibres are difficult to
make in high strength form, probably because of structural
defects arising from shrinkage during firing and drying, but
values of Young's modulus of 400 GN m^{-2} and strengths of
about 2,000 MN m^{-2} have been reported (Sutton, 1968).

C.   GLASS AND SILICA FIBRES

One of the most useful forms of high-strength oxides
is glass fibre. Ordinary glasses and glasses for fibre
manufacturing do not have unique compositions and the
commonest form is E-glass. This is a lime-alumina-
borosilicate glass based on the eutectic at the composition
SiO_2(62%) Al_2O_3(14.7%) CaO(23.3%). E-glass is a low alkali
glass, with some boric oxide substituted for SiO_2 to act as
a network former, and magnesia is sometimes substituted for
lime. A more recently developed glass, S-glass, is stronger
and more expensive than E-glass, and contains silica, alumina
and magnesia in the weight ratio 65 : 25 : 10.

In the manufacture of fibre, glass marbles are melted
and exuded through platinum nozzles. The fibres are then
drawn rapidly and wound onto a drum. Some 200 fibres

between 10 and 20 microns in diameter are drawn together to
form a tow.   During drawing the rate of cooling is very
high and crystallisation is suppressed.   The glass structure is a disordered network built up from units in which
each silicon atom is tetrahedrally-bonded to four oxygen
atoms.   During drawing, surface or volume defects which
were present in the bulk glass are eliminated and freshly-
drawn fibres are therefore extremely strong.   In this state
they are susceptible to damage through handling or abrasion
and must be protected by the application of a varnish.   The
average strength of commercial E-glass fibre is between
1,500 and 2,000 MN m^{-2}, but with care in preparation and
handling strengths of 3,700 MN m^{-2} have been obtained.
S-glass is 20% stronger than E-glass, while fibres of high-
purity silica have measured strengths of 7,000 MN m^{-2}.
Early experiments with glass fibres showed a strong dependence of strength on fibre diameter, but Thomas (1960) has
shown that the short-time breaking strength of carefully
prepared E-glass fibres between 5 and 50 microns in diameter
is constant.   The high strength of fresh fibre is reduced
by heating, and by exposure to moisture.   All glasses are
to some extent attacked by water.

IV.   STRONG BUT BRITTLE SOLIDS

Table I shows typical values of the strength and
modulus of alumina and glasses.   These can be compared in
terms of the fraction of the theoretical strength attained
(Fig.3).   Conventional hot-pressed ceramic bodies reach
about 1% of the theoretical strength.   Single crystals are
little better unless their surfaces are flame-polished, when
strength may be as high as 10% of the ideal value.   The

strength of most whiskers is not generally much higher.

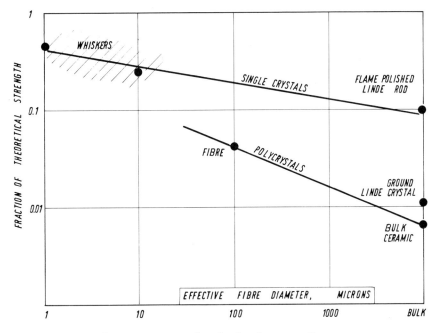

Fig.3. Strength of alumina samples.

In the bulk form the strength of glassy solids is three orders of magnitude less than the theoretical fracture stress, but in fibre form strengths as high as 10-50% of the theoretical value are common.   When the network structure of a glass is broken it is the covalent bonds that determine the strength.   But networks are not rigid, and they can be deformed easily by quite small forces.   As a result, glasses have low elastic moduli, a feature that has limited wide use of glass fibre.   Alumina, and some of the newer fibres such as carbon and boron, are five to ten times as rigid as glass.   In terms of cost and strength-to-cost ratio, however, glass fibres have advantages over sapphire whiskers.

Alumina and silica have no mechanism by which gross plastic deformation can take place at ordinary temperatures. Any defect present either in the surface or in the interior will consequently act as a source of stress concentration, which cannot be relieved by plastic shearing and the defect will propagate as a brittle crack.   Because the distribution of flaw sizes is usually quite random, the failure of a brittle solid is unpredictable.   The empirical Weibull relationship, which is often used for design purposes, gives the probability of failure, P, as

$$P = 1 - \exp \left[\frac{V(\sigma - \sigma_u)}{\sigma_o} \right]^m$$

where σ is the applied stress, σ_u is the stress at which there is zero probability of failure, σ_o is a normalising factor, V is the volume of material, and m = material flaw density constant.

A low value of m indicates a material with flaws of variable severity, while a high m indicates a material with a uniform distribution of regular flaws.   For design purposes materials with high values of σ_u and m are preferred. Table II shows experimental values of m for some common materials.   In some cases a brittle solid has a value of $\sigma_u \to$ zero.   This implies that despite the potential strength of the solid, it would not be safe to design on the basis of tensile stresses in such a material.   On this account, alumina and glass might never be used for major applications in the bulk state.   However, if the brittle solid is used in particulate or fibrous form, and is

embedded in a matrix of a material with a higher degree of
tolerance for cracks, then failure of some pieces under low
loads will not result in catastrophic failure of the whole
composite.

TABLE II

Material	m
Glass fibres	1.8
Cold-pressed and sintered Al_2O_3*	3.25
SiC	4.2
Ground Al_2O_3*	6.0
Nickel-bonded TiC	7.0
Al_2O_3 after annealing at 1200^oC*	11.0
Graphite	12.0
Al_2O_3 ground and then annealed at 1200^oC	12.25
Structural steel at - 200^oC	24.0
Spark-plug porcelain	35.0
Steel at R.T.	58.0
Perfect Material	∞

* Shows the effect of a sequence of post-manufacturing
operations on one batch of Al_2O_3

V. PARTICULATE COMPOSITES

The character of a composite is largely defined by the
volume fraction, V_f, and by the dimensions of the non-
continuous phase in terms of the aspect ratio, l/d. For a
composite containing continuous fibres, l/d = ∞, while for a
particulate composite, l/d = 1. For non-continuous fibres,
the aspect ratio falls between these extremes, while the
important parameter is the particle size in particulate
composites. Oxide particles can be ground to 0.1μ diameter

without difficulty, and small volume fractions can be mixed
directly with a metal matrix.   Alternatively, the oxide may
be introduced by reaction, e.g. aluminium or silicon with
oxygen, both the metal and the gas being in solid solution
in a non-reactive metal such as copper.   The oxide particles
grown in this way may be less than 0.01μ in diameter.
Finally, a compacted but porous aggregate of oxide particles
may be infiltrated with molten metal.   The properties of
these three types of material are very different, because of
their distinctive strengthening mechanisms.   A uniform
distribution of fine particles is effective in blocking
dislocations; a material containing such a distribution has
a high work-hardening rate and rapidly develops a dense and
stable dislocation network.   Sintered aluminium powder
(SAP), and TD nickel (nickel containing about 2% ThO_2
particles in very fine dispersion) are alloys which depend
for their high-temperature strength upon this form of dislocation sub-structure built up by an extended series of
thermal and mechanical treatments.   The ceramic content in
these materials is usually much less than 10 vol.%.

     Materials consisting of ceramic particles separated by
only a small amount of metal do not rely on dislocation
hardening mechanisms.   The metal component, which is still
usually the continuous phase, separates the oxide particles
so that cracks developed in them should not propagate
rapidly through the whole body.   The strength and modulus
of the composite are usually less than those of the ceramic
phase itself, but the metal matrix increases the work of
fracture by plastic blunting of the tips of cracks that may
have originated in, or passed through, the ceramic particles.
This type of composite is known as a cermet, and usually

contains between 20 and 70 vol.% of ceramic phase.

Cermets can be manufactured by most of the standard techniques for forming ceramic materials.   Graded ceramic/ metal powder mixes are cold-pressed and sintered, hot-pressed, extruded in water- or polymer-based pastes, or slip-cast as slurries.   The final sintering operation causes the metallic binder to fill all the voids, and promotes good adhesion by encouraging slight alloying between ceramic and matrix.   For many years cemented carbides have been extensively used for high-temperature components, and their usefulness is due to the strong bond that exists between the carbide particles and the nickel or cobalt binder.   The strengths of these materials can be 1,000 or 2,000 MN m^{-2}, and for a typical composite containing 50 wt.% of WC in Co, Young's modulus is of the order of 350 GN m^{-2}. The elastic modulus frequently obeys a simple form of mixture rule as a function of composition.

Unlike cemented carbides, bonding between phases in oxide-based cermets is usually poor and in consequence it is necessary to improve the wetting of the ceramic by the metal in order to obtain high quality composites with low porosity and high strength.   The best known of the oxide cermets is the Cr/Al_2O_3 system.   In this system there is no reaction between the phases unless a small amount of Cr_2O_3 is present, and the compacts are therefore usually sintered in an oxidising atmosphere.   The film of Cr_2O_3 which forms on the metal is isomorphous with Al_2O_3 and is able to dissolve slightly in the alumina and so increase wetting.   In iron, cobalt, and nickel-bonded alumina cermets a spinel phase forms which serves the same purpose.   Approximately 7% of Cr_2O_3 is needed in Cr/Al_2O_3 cermets if a strong and

non-porous composite is to be obtained.   A small quantity
of titanium, which has a greater affinity for oxides than
nickel, is sometimes added to achieve the same purpose.
One of the basic compositions in the Cr/Al_2O_3 system contains 70 wt.% of ceramic.   As Table III indicates, this
material has low strength but a high elastic modulus.   It
has good resistance to oxidation and fair resistance to
thermal shock at temperatures of $\sim$ 1200°C.   Cermets with
only 28% of alumina have better thermal shock resistance,
and roughly the same mechanical properties, but high strength
and good thermal shock resistance can be achieved by using
an 80/20 Cr/Mo alloy as matrix.   Oxide-based cermets do not
compare very favourably with cemented carbides.   Their most
serious defect is poor fracture toughness.

TABLE III

Material	Density (10^3 Kg m^{-3})	Tensile Strength (MN m^{-2})	Young's Modulus (GN m^{-2})	Charpy Impact Energy on Standard Un-notched Samples* (KJ m^{-2})
70% Al_2O_3/ 30% Cr	4.7	245	365	< 5
28% Al_2O_3/ 72% Cr	5.9	275	330	< 5
34% Al_2O_3/ 66% CrMo (80:20) alloy	5.8	371	310	$\sim$ 5

*Reported as Charpy impact energy (W) divided by twice the
specimen cross-sectional area (2A) to obtain units comparable
with work of fracture values quoted elsewhere (γ_F).

   Oxide cermets have been considered for service in
applications where their high thermal stability, coupled
with thermal shock resistance and ease of forming, can best

be exploited.   They have been used, for example, as rocket
nozzle inserts where there is a serious hot gas erosion
problem.   Their considerable resistance to erosion by
molten metals has enabled them to be used as flow control
valves, thermocouple tubes, and pouring spouts in the metallurgical industry.   They have been used as mechanical seals
and bearings for service under conditions of poor lubrication
at high temperatures and in aggressive chemical environments
like hot, concentrated sulphuric acid.   Whilst they cannot
be used as gas turbine blading because of their poor impact
resistance, they have been used as flame spreaders in jet
engine combustion chambers.   With a small amount of Ti as a
binder, Al_2O_3 becomes a useful cutting tool material for
hard steels and alloys.   It is more brittle than carbide
tools but has greater wear resistance and is being increasingly used under conditions where there is little mechanical
shock.

One of the most successful applications of oxide/metal
composites has been in the nuclear power industry. ThO_2
and UO_2 fuels are frequently dispersed in finely-divided
form in an aluminium or stainless steel matrix for the
fabrication of fuel elements, a device which permits a high
degree of fuel burn-up without destroying the mechanical
integrity of the fuel element.

Finally, mention is made of two rather more familiar
oxide-containing composites, porcelain and concrete.
Porcelain is a glass-bonded ceramic containing a mixture of
crystalline quartz and mullite ($3Al_2O_3.SiO_2$).   The variety
of available compositions and mechanical properties is wide,
but strengths of porcelains are usually between 80 and 160
MN m^{-2} and their elastic moduli are of the order of 100 to

200 GN m^{-2}. Toughness is hard to assess and is generally low, but as Table II shows the flaw density parameter, m, for hard porcelains is intermediate between that of brittle and ductile materials, which implies a relatively high toughness. The high value of m may be due to the presence of a surface glaze.

Concrete is also a complex material containing SiO_2 (in the form of sand particles) or Al_2O_3. In a typical mortar, sand particles are embedded in a cement gel matrix which also contains unhydrated cement particles. The cement is largely composed of anhydrous calcium silicates and aluminates which on hydration form a finely-crystalline gel. The hydration process continues slowly with time, and the compressive strengths of mortars have been found to be still increasing even 50 years after manufacture. In addition to the aggregate and binder, most concrete mixes contain voids and water. The voids are a result of the gradual absorption into the gel of water, and the strength of concrete is therefore highest when the initial water/cement ratio is kept to the minimum consistent with adequate workability. The mechanical properties of such a complex system depend on many variables, including the volume fraction and mechanical properties of the additives, but tensile strengths as high as 7 MN m^{-2} and elastic moduli of the order of 35 GN m^{-2} are not uncommon. The fracture toughness of concrete may be as low as 10 to 200 J m^{-2} for 1 : 3 cement/sand mortars, but these low values can be increased an order of magnitude by incorporating steel or glass fibres into the mixes.

VI.   FIBRE-REINFORCED COMPOSITES

   Table I gives an indication of the gains to be made,
in terms of composite strength, by using fibres instead of
particles as reinforcing agents.   In cermets, strengths
are usually less than five times the matrix strength,
whereas, in theory, composites could be made which are fifty
times stronger than the matrix.   The mechanism of fibre
strengthening is quite different from that in cermets.   A
load applied to a composite is shared between the fibre and
the matrix and provided they are constrained to deform
together, i.e. well-bonded, the total deformation of the
composite is less than that of the matrix and the modulus is
higher.   A bundle of fibres has a breaking strength lower
than the average strength of fibres in the bundle, and its
strength is less sensitive to the presence of cracks than
that of a piece of bulk material of the same shape.   When
the bundle is embedded in a matrix, the matrix fulfils
several functions.   It protects fibre surfaces from damage
and loss of strength;   it separates fibres so that cracks
cannot pass from one to another, and it provides a means of
transferring load into the fibres.

   In a composite containing more than a few volume percent (V_{min}) of continuous, aligned fibres the tensile
strength is often given by a simple rule of mixtures

$$\sigma_c = \sigma_f V_f + \sigma'_m (1 - V_f) \ldots \ldots (V_f > V_{min})$$

where σ'_m is the stress in the matrix when the composite is
strained to its ultimate tensile strength.   When the fibres
are not continuous, stress must be transferred into each
fibre by shear along the fibre/matrix interface.   A certain

length of each fibre is therefore not fully utilised, and
the composite strength is given by the relationship

$$\sigma_c = \sigma_f V_f \{1 - (1 - \beta) \, l_c/l\} + \sigma'_m(1 - V_f)...(V_f > V_{min})$$

The critical transfer length, l_c, represents that length of
fibre which may be stressed to its breaking strength within
the composite, and β is determined by the manner in which
the stress builds up from the ends of the fibre.   Short
fibres will thus always strengthen less than continuous
ones, but even if the fibres are only $10 \times l_c$ long, 95% of the
strength available with continuous fibres can be obtained
with discontinuous ones.   Fig.4 (Sutton and Chorne, 1964)
shows the combined effects of volume fraction, fibre
strength, and aspect ratio on the strength of the composite.
In order to make the most of any short fibre, stress
transfer must be as efficient as possible, and the nature of
the interfacial bond between fibre and matrix is therefore a
matter of prime importance.   A considerable amount of
research has been expended in developing coatings for both
whisker and glass surfaces to promote good wetting or a high
interfacial bond strength.

So far, it has been assumed that the fibres, long or
short, are aligned parallel with the direction of the
applied stress.   In such composites the mechanical properties are highly anisotropic.   Both strength and elastic
modulus fall off very rapidly when the stress axis deviates
by more than 5 to $10°$ from the fibre axis.   This is
reflected in the strength of short fibre composites because
of the difficulty in ensuring that all fibres in the composite are exactly aligned.

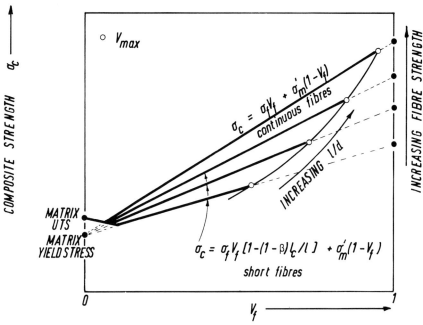

Fig.4. The effects of fibre strength, fibre volume fraction, and fibre length on the strength of composites.

The fracture toughness of a fibre composite is a much more complicated property to assess than that of particulate composites. Many fibre-reinforced materials are much tougher than might be expected from the toughness of the constituents. This is largely a result of the work required to pull out of the cracked matrix the ends of fibres that have broken some distance away from the plane of the crack, as illustrated in Fig.5. This may occur if, as a result of the lateral stress concentration ahead of a spreading crack, fibres and resin are debonded before the crack tip reaches them (Cook and Gordon, 1964).

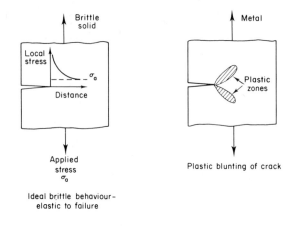

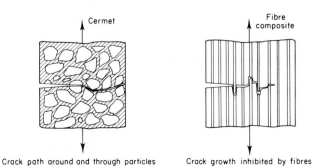

Fig.5. Schematic illustration of the fracture process in brittle solids, metals, cermets and fibre composites.

Cottrell (1964) has estimated that when pull-out occurs the work of a composite, γ_F, is proportional to $V_f \sigma_f l_c$. For high toughness the requirements are a high fibre strength and a long critical length.   The critical aspect ratio can be determined by equating the shear forces at the fibre/resin interface with the tensile force in the fibre at the moment of fracture, and is given by

$$l_c/d = \sigma_f/2\tau$$

where τ may represent the fibre/matrix interfacial friction
stress or the matrix shear yield stress, whichever is the
smaller.   The fracture toughness is then

$$\gamma_F = \frac{V_f \tau l_c^2}{12d}$$

The work of pulling out fibre ends is thus related to the
strength of the interfacial bond and to the square of the
pull-out length.   As a consequence the toughness of a fibre
composite will increase with increasing bond strength when
the bond is still very weak, but will pass through a maximum
and fall with further increases in bond strength.   Cooper
and Kelly (1967) have also shown that toughness increases
with the diameter of the reinforcing fibre, a good reason
for using large diameter, continuous ceramic fibres instead
of whiskers.

A.   MANUFACTURE OF FIBRE REINFORCED COMPOSITES

Metal matrix composites containing continuous glass or
silica fibres have largely been manufactured by hot-pressing
techniques, although a considerable amount of work has also
been carried out on liquid infiltration methods and on
electroforming combined with filament-winding.   It is more
difficult to make whisker-reinforced metals because of the
problems of aligning the very fine crystals.   Extrusion of
mixtures of whiskers and fine metal powder usually results
in whisker breakage, and may be unsatisfactory even though
it results in good compaction and matrix properties.   If
whiskers and metal powder are dispersed in ammonium alginate
solution and extruded into an acid bath, the alginate gels
and the fibres are held in a reasonably well-aligned state.

The alginate can be burnt off and the resulting whisker/
metal frits hot-pressed or liquid-phase sintered to give
dense composites.   High volume fractions of well-aligned
whiskers or short fibres are difficult to obtain.

Glass/resin systems are more common.   Continuous
fibre can be incorporated into resins by filament-winding,
or by pre-impregnation and hot-pressing (dry lay-up).
Chopped fibre can be incorporated by hand lay-up with liquid
resin or it may be used in dough-moulding compounds in which
it is pre-compounded with resin, chalk filler, and a thermo-
plastic phase which mitigates shrinkage stresses.   Glass
reinforced thermoplastics are made by injection moulding,
but there are problems in maintaining fibre lengths longer
than the critical value, and careful feeder design is
necessary to avoid excessive fibre breakage.

B.   PROPERTIES OF REINFORCED METALS

Research in this field has been concentrated almost
exclusively on the preparation and properties of metals such
as aluminium, silver and copper reinforced with sapphire
whiskers and glass fibres, although other metal and alloy
matrices have also been used.   Silver and nichrome can be
strengthened by sapphire whiskers and relatively large
increases in strength can be obtained by the introduction of
small fractions of whiskers, provided the matrix can be made
to wet the whiskers.   Fig.6 compares the properties of
aluminium reinforced with glass, silica, and Al_2O_3 whiskers.
The strengths of glass and silica reinforced aluminium are
much lower than anticipated from the mixture rule, possibly
due to fibre damage during manufacture.   As fibres are
frequently more stable than either work-hardened or

precipitation-hardened structures in low melting point
metals and alloys, the strength of fibre composites can
often be retained at much higher temperatures. Al_2O_3
whiskers remain strong at temperatures up to the melting
point of silver, but glass and silica, being viscous solids,
lose their strength much more rapidly.   At higher temperatures and in other matrices, especially those containing
nickel, Al_2O_3 whiskers tend to lose strength due to a change
in morphology.

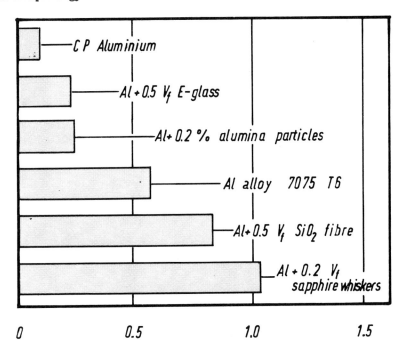

STRENGTH GNm^{-2}

*Fig.6.   Room-temperature strength of aluminium
reinforced with glass and with whiskers,
compared with the strengths of pure aluminium
and fully-heated 7075 alloy.*

C.   PROPERTIES OF REINFORCED PLASTICS

The data on glass-fibre reinforced plastics (GRP) is
too great to discuss in detail here.   Polyester and epoxy
resins and a variety of thermoplastics are commonly
reinforced both with continuous and chopped glass fibre, and
many are used on a large scale.   Although most commercial
GRP products do not have particularly high strength, a high
proportion of the available strength of the glass can be
retained by filament winding methods in which almost perfect
fibre alignment is possible.   Consequently, although some
experiments have been carried out on whisker-reinforced
resins, there is no real incentive, in the way of a large
gain in strength-to-weight ratio, to encourage such work.
GRP's have two main disadvantages, namely a relatively low
stiffness and weakening when exposed to moisture.   The
latter effect can be inhibited by treating the glass surface
with suitable coupling agents.   Although this offers scope
for whisker reinforcement, asbestos or carbon fibres are
more likely to be used to add rigidity.

D.   TOUGHNESS OF FIBRE-REINFORCED COMPOSITES

Most metals are capable of some degree of plastic
deformation to ensure that cracks or notches can be blunted,
and the metal is consequently tough.   The embedding of
brittle fibres of glass, silica, or alumina in a ductile
metal reduces the fracture strain and toughness of the metal.
A similar effect occurs in particulate composites.   Hence,
in these systems any gains in strength and modulus will
always entail a loss of toughness.   However, the presence
of fibres, even brittle ones, frequently makes an alloy
insensitive to the presence of notches.

In brittle matrices the effect of fibres on toughness
is quite remarkable, as illustrated by glass-fibre reinforced
epoxy or polyester resin.   The fracture energy of glass is
$\sim$ 10 J m^{-2} and that of the resin 200 to 500 J m^{-2}.   However,
a composite containing 70 vol.% of aligned E-glass fibres
has a toughness of nearly 2×10^5 J m^{-2}, which is clearly not
the sum of the separate energies of formation of new fracture surfaces.   When such composites are fractured, the
fibre ends that are pulled out of the resin are about a
millimetre long.   Increasing the fibre/resin bond strength
reduces the fibre pull-out length and severely reduces the
fracture toughness.   The beneficial effect of an increase
in toughness when two brittle materials are combined has
been used in attempts to reinforce bulk silicon nitride with
silicon carbide whiskers and to reinforce glasses and silica
with carbon fibres.   The toughness of a range of composites
is compared with that of more familiar homogeneous materials
in Fig.7.

E.   FIBRE-REINFORCED CERAMICS
The problem of manufacturing composites in which the
continuous phase is a ceramic material is more troublesome
than that of reinforcing resins, which are never quite as
brittle.   The greatest difficulty lies in the thermal
expansion mismatch which results in large residual stresses
when a ceramic/metal composite is cooled from its pressing
temperature.   The resulting composites are therefore often
cracked and of low strength.   Experiments have been carried
out on the reinforcement of thoria, alumina and clay bodies
with molybdenum and tungsten wires, and these composites
have better shock resistance than the unreinforced matrix,

despite the presence of cracks and the lower strengths. In
the mullite/molybdenum system the thermal expansion coefficients are nearly the same and it is possible to obtain
high strengths and uncracked composites with useful
properties. The problem with refractory metal fibres,
however, is that they oxidise and degrade during pressing or
subsequent service, and stainless steel or nichrome wires
are therefore preferred. The upper limit of service temperature is much lower with fibres of this type.

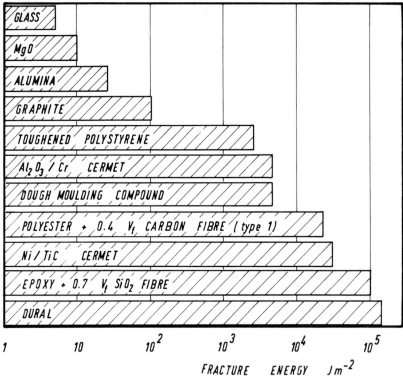

Fig.7. Comparison of the toughness
of various engineering materials.

REFERENCES

Bayer, P.D. and Cooper, R.E. (1967) J. Mat. Sci. 2, 333.

Cook, J. and Gordon, J.E. (1964) Proc. Roy. Soc. A282, 508.

Cooper, G.A. and Kelly, A. (1967) J. Mech. Phys. Solids
  15, 279.

Cottrell, A.H. (1964) Proc. Roy. Soc. A282, 508.

Kelly, A. (1966) "Strong Solids", Clarendon Press, Oxford.

Olds, G.C.E. (1966) "New Ceramics", Science Journal,
  August, 58.

Sutton, W.H. (1968) "Ceramic and Graphite Fibres and
  Whiskers" Vol.3 (L.R. McCreight, H.W. Rauch and W.H. Sutton,
  eds.) Academic Press.

Sutton, W.H. and Chorne, J. (1964) "Fibre Composite
  Materials", A.S.M., Cleveland.

Thomas, W.F. (1960) Phys. Chem. Glass 1, 4.

Webb, W.W. and Forgeng, W.D. (1957) J. Appl. Physics 28,
  1449.

GENERAL REFERENCES AND DATA SOURCES

"Ceramic and Graphite Fibres and Whiskers" (1965) Vol.1 and
  (1968) Vol.3 (L.R. McCreight, H.W. Rauch and W.H. Sutton,
  eds.) Academic Press.

"Whisker Technology" (1970) (A.P. Levitt, ed.) Wiley-
  Interscience.

"Cermets" (1960) (J.R. Tinkelpaugh and W.B. Crandall, eds.)
  Reinhold, New York.

"High Temperature Materials and Technology" (1967)
  (I.E. Campbell and E.M. Sherwood, eds.) Wiley, New York.

CHEMICAL PREPARATION AND ANALYSIS OF OXIDES

N.J. BRADLEY, B.S. COOPER AND D.J. HOBBS

Johnson Matthey Chemicals Limited,
Royston, Hertfordshire

I.   INTRODUCTION

II. PRODUCTION

III. PREPARATION OF OXIDE MATERIALS

    A. ALUMINIUM OXIDE

    B. SILICON DIOXIDE

    C. LEAD DIOXIDE

    D. GALLIUM OXIDE

    E. BORIC OXIDE

    F. NIOBIUM PENTOXIDE

    G. FERRIC OXIDE

    H. RARE-EARTHS AND RELATED OXIDES

    I. BISMUTH GERMANATE

IV. CHARACTERISATION OF OXIDES

    A. MAJOR CONSTITUENTS

    B. TRACE CONSTITUENTS

    C. SPECTROGRAPHIC TECHNIQUE

I.   INTRODUCTION

Pure oxides are used in the preparation of lasers,
electro-optic devices, ferrites and for numerous other

applications. In this article, examples have been chosen
to illustrate the wide scope of the demands of the electronics industry, and the necessary stringent restrictions
this places on their production. Details of specific
methods of manufacture will be given together with more
general purification procedures, and details of analytical
techniques available for process control.

In many cases the purity of a product cannot be
guaranteed by starting with good materials, but stringent
analytical control markedly improves this situation.
Controls include wet chemical analysis, particle size
assessment, density measurement, X-ray diffraction studies
and conventional spectrographic analysis.

Compositional changes in the starting material can
result in differences in the end-product unless they are
detected by analysis and the process modified accordingly.
Analysis of the final product is essential to ensure that
it is both stoichiometric and pure. In many instances,
unless specialist analytical instrumentation is available,
'device performance' is the ultimate measure of purity.
Feedback of performance data is essential to the manu-
facturer and enables high purity materials to be continually
improved.

II. PRODUCTION

Very few materials occur in nature in a purity or form
suitable for device application and in consequence chemical
and physical processing is essential. The preparative
route used for a specific oxide material is determined by
two criteria; firstly, the target specification of the
required oxide and secondly, the available starting material.

Special steps must be taken to remove elements deleterious to device performance, which requires a target purity in parts per million.   It is essential that the user must receive not only a pure product, but also one having the correct composition, stoichiometry and form (α, β, γ, etc.).   The material may be required as a fine or dense powder with a specific particle size, or as fused lumps or broken polycrystalline pieces of a particular density or size.

In most cases the starting materials are chosen to produce the required material at the lowest cost.   Hence a compromise must be obtained between intrinsic material cost and the amount of processing required to upgrade the starting material.   In the preparation of many oxides, and particularly those containing rarer elements, the choice of starting materials is limited to a mineral concentrate. For common elements it is usual to start with a technical grade salt or metal.

Purification procedures play a large part in the preparative chemistry of oxide materials.   There are no universal or absolute purification methods, and the whole practice of reducing impurities or refining a substance is a relative one.   There are seven basic methods of purification in current use.   Four classical methods have been in use for over a hundred years;   these are distillation, crystallisation, precipitation and electrolysis.   Three newer techniques, which are widely used in large scale preparations are ion exchange, solvent extraction and zone-refining.

In the preparation of any one specific high purity material it is common to use a combination of one or more

of these techniques.   Each technique may need to be
repeated several times to achieve the required purity in
the refined material, particularly in solvent extraction
and crystallisation procedures.   Consequently, after each
cycle there is a rejection of fractions in which the
impurity concentration has increased.   A characteristic of
some purification procedures is the low yield efficiency
which is a major cost factor.

The scale of operation for most high purity oxides is
10 - 500Kg.   Batch processes are common but continuous
processes are also used.   Most processes for ultra pure
materials have a primary efficiency or yield of 5 - 20%.
The rejected fraction is discarded or used in processes
where reagent grade material is acceptable.   For the rare
and precious elements, the reject fraction is returned to a
secondary refining circuit for upgrading.   Multiple purification methods follow the law of diminishing return as
illustrated in Fig.1.

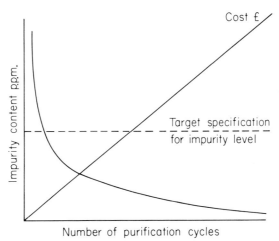

*Fig.1.   Impurity content per number
of purification cycles.*

Ideally a purification procedure is chosen in which the rate of change of impurity concentration is large. However, the target specifications are often beyond the practical limit which is set by: (1) the equilibrium point of the purification process, (2) contamination of the product from impurities in the processing reagents and (3) contamination from the materials used to construct the process plant.

All large scale preparations and packaging should be carried out in ultra clean conditions to limit cross contamination during processing.

III. PREPARATION OF OXIDE MATERIALS

A. ALUMINIUM OXIDE

Aluminium oxide (Al_2O_3) is used in the manufacture of fibre optic glasses, lasers, insulating substrates and ceramic oxides. High density material free from transition metal ions is required for crystal growing applications, and a low density form is used for the preparation of blends and mixed oxides.

Alumina is used in conjunction with yttrium oxide in the preparation of yttrium aluminium garnet (YAG) and yttrium aluminium perovskite (YAP) for laser applications and with magnesia for the preparation of spinels ($MgAl_2O_4$ etc.) and insulating substrates.

The material can be readily prepared from commercial grade aluminium metal in which the main impurities are iron and sodium. The metal is dissolved in concentrated hydrochloric acid in the presence of mercury. Mercury dissolves oxide scum which would normally inhibit the reaction. Excess aluminium metal is used to prevent the mercury from dissolving, and the aluminium chloride solution is separated. The aluminium chloride is recrystallised numerous times to increase the purity and the product is roasted in air, which removes any residual mercury.

A high density material can be prepared either by calcination or by isostatic pressing. A low density form can be obtained by the careful decomposition of aluminium hydroxide or oxalate.

B.   SILICON DIOXIDE

Silica (SiO_2) is widely used in the preparation of synthetic quartz, for X-ray windows, and silicates for use as phosphor materials, glasses and in fibre optic applications. This material is also extensively used in microelectronics and as a masking material during device manufacture, though for some applications it has been superseded by silicon nitride.

An extremely high purity grade of silica can be obtained by burning zone-refined silicon in chlorine below $400^{\circ}C$. The silicon tetrachloride is purified by distillation and subsequently hydrolysed to give a finely divided suspension of silicon dioxide which can be separated and fired. A much cheaper technique is to fractionally distil a by-product of organo-silane manufacture which can then be used as a starting material.

C.   LEAD DIOXIDE

Lead dioxide (PbO_2) has widespread uses in the preparation of glasses, notably low-loss fibre optic glasses for telecommunications. It has also been used as a flux component in the preparation of single crystal material such as rare-earth orthoferrites.   In conjunction with molybdenum trioxide, it is used in the manufacture of lead molybdate for acousto-optics.

For these applications the lead oxide should be ultra pure, and low in transition metal impurities (<1 p.p.m.). Ideally it should not contain any free lead which can attack the platinum vessels used in the preparation of glasses and in crystal growth techniques.

Lead dioxide can be readily prepared by electrolysis of a very high purity lead nitrate solution using platinum electrodes.   The pure lead nitrate is obtained by multiple crystallisation techniques.

Purer grades of material can be obtained by careful control of the pH during electrolysis and by discarding the first and last fractions of the oxide obtained.   After recovery from the electrode the lead dioxide is washed in dilute nitric acid and water to remove traces of calcium and other soluble salts.   Roasting in oxygen at 160°C for several hours removes any trace of water.

It is possible to test for the presence of free lead by preparing a small quantity of lead glass in a platinum crucible and observing any attack.   The transparent appearance of the glass, both when molten and at room temperature, is an extremely useful indication of the quality of the lead dioxide.

D.   GALLIUM OXIDE

Gallium oxide (Ga_2O_3) has numerous applications in the preparation of ferroelectric and piezoelectric materials. Gallates also have extensive use as phosphors in cathode ray tubes for commercial and military uses.

The oxide can be prepared from gallium metal extracted from the enriched ores which are a by-product of aluminium or zinc smelting. An ultra high purity material can be obtained by starting with gallium metal having a known residual resistivity ratio.

The metal can be purified by dissolution in hydrochloric acid followed by treatment with sulphuric acid, hydrogen sulphide and diethyl ether to remove insoluble sulphates, sulphides and iron. To prepare the oxide, excess oxalic acid is added to the solution to give a fine precipitate which is filtered, washed and fired above 600°C to give the β form.

E.   BORIC OXIDE

Boric oxide (B_2O_3) is widely used in crystal growing applications both as a flux component and as an encapsulant. It forms a heavy viscous liquid which has a high chemical stability and low vapour pressure. Boric oxide will remove any metallic oxide scum and can be used for controlled doping with volatile materials. It is also used in the preparation of microelectronic devices and glass manufacture.

Boric acid occurs naturally, with traces of magnesia and organics, in the salt deposits around hot springs. The most effective purification process is multiple crystallisation from water. The product is then roasted, after treatment with nitric acid, to oxidise the volatile organic impurities.

The material can be prepared either as a powder or as fused lumps. For crystal growing applications it is cast into transparent discs of any shape or size. The material is extremely hygroscopic and due to its great affinity for water special care must be taken in handling and packaging. For crystal growth applications, the fused material must be clear, transparent and free from inclusions and air bubbles.

F.   NIOBIUM PENTOXIDE

Niobium pentoxide (Nb_2O_5) is extensively used for the preparation of the niobates mainly for electro-optic applications, e.g. lithium niobate.   Complex niobates such as barium sodium niobate and potassium tantalo-niobate, also used in electro-optics, can be obtained by roasting the carbonates of the potassium, sodium, lithium, strontium, magnesium, barium, lead, tantalum, etc. with Nb_2O_5.

Niobia can be extracted from niobium rich ores, normally containing tantalum, iron, titanium and tin as the principal impurities, by either ion exchange or solvent extraction.   In solvent extraction the ores are slurried in hydrofluoric and sulphuric acid and the niobium rich salts extracted with methyl isobutyl ketone or cyclohexane. Excess H_2SO_4/HF is then added, and the niobium is finally precipitated with ammonia.   The filtrate is washed, and fired at 1200°C to give a high purity Nb_2O_5.

A high density polycrystalline product of some of the niobates can be obtained by careful pulling from the melt, by the Kyropoulos technique.   Slight colouration of the polycrystalline material can be removed by annealing in oxygen.   The mixed niobates, however, are readily formed by making a slurry of the metal carbonates and Nb_2O_5 and firing at 1200°C.   The product, when cool, is washed in dilute nitric acid which reacts with any final traces of carbonate. The mixture is then refired to give a homogeneous product.

G.   FERRIC OXIDE

Ferric oxide (Fe_2O_3) is widely used in the preparation
of magnetic oxides, orthoferrites and garnets for bubble
domain memory devices.   Along with yttria, it has extensive
microwave applications in the form of yttrium iron garnet
(YIG).

The oxide can be prepared by dissolving iron metal in
hydrochloric acid to give ferrous chloride.   This is
refluxed and ammonium polysulphide is added to precipitate
the main impurities as insoluble sulphides which are removed
by filtration.   The iron can be precipitated as the oxalate
which, after firing, gives the high purity red ferric oxide.

H.   RARE-EARTHS AND RELATED OXIDES

These oxides are extensively used in the preparation
of ceramics, pigments, luminescent materials and phosphors
for both cathode ray tube and military applications.   The
rare-earths also have applications in the preparation of
garnets for lasers and magnetic ferrites.

The oxides can be extracted from enriched ores,
normally a by-product of uranium mining, by ion exchange.
The ores are nitrated and passed into ion exchange columns,
where a reaction occurs between the rare-earths and the
cationic insoluble acid used as a resin.   When a sufficient
quantity of material has been allowed to collect in the
columns, the system is elutriated with E.D.T.A. (ethyl
diamine tetroacetic acid).   This returns the resin to its
original state ready for re-use and releases the rare-earths
as chelate.   The heavy fraction is the first to wash out
(Lu, Yb, Tm, etc. down to Ce) and the product is collected
in a fractional collector and washed with dilute nitric acid.

The rare-earth is then precipitated with excess oxalic acid, and the product roasted to give a high purity oxide.

Careful control or repetition of this process can lead to very high purity rare-earth oxides having particularly low concentrations of other rare-earths.

I.   BISMUTH GERMANATE

The compound bismuth germanate ($Bi_{12}GeO_{20}$) has recently found application in microwave acoustic delay line systems. This mixed oxide is normally prepared by co-precipitating the oxides of germanium and bismuth in the required proportions, followed by firing at $1000^{\circ}C$.

High purity germanium dioxide which is extensively used for germanate phosphor production and in the preparation of modulators, can be prepared by dissolving germanium metal in hydrochloric acid.   The resulting germanium tetrachloride can be distilled to give a very pure product which, on hydrolysis, followed by filtration and firing, gives the oxide.

High purity bismuth oxide, sometimes used as a flux additive in crystal growth to lower the melt viscosity, is normally prepared by dissolving bismuth metal in nitric acid. The nitrate can then be purified by multiple crystallisations and will give a fine oxide powder when fired.

Two preparative techniques can be used to great advantage on the mixed oxide systems.   The first is the process of co-precipitation in which homogeneous mixtures can be obtained by precipitation of the hydroxides or oxalates (if insoluble) from a solution containing the elements as soluble chlorides or nitrates.   The filtrate is then fired to give a homogeneous mixed oxide.   This is ideal for the preparation of dilute mixtures or for non-stoichiometric compositions, and is extensively used in the manufacture of phosphors and materials such as lead titanate.

The second technique involves the use of organo-metallics. In the case of $Bi_{12}GeO_{20}$ a mixture of bismuth oxalate and germanium alkoxides or tetraethylorthogermanate $Ge(OC_2H_5)_4$ can be hydrolysed and the filtrate fired to give a homogeneous mixture.

The principal advantage of this type of preparation is that the organometallic compounds can be fractionally distilled to extremely high levels of purity, and on roasting all volatile organic components are removed. Numerous suit-able organometallic compounds are available (e.g. aluminium isopropoxide) which lend themselves readily to the prepara-tion of dilute mixed oxide systems and are available for use as dopants in vapour epitaxy systems (e.g. trimethyl gallium).

Many other sophisticated techniques are available, such as microbiological and chromatographic separations, but so far these seem to be of limited practical application.

IV.   CHARACTERISATION OF OXIDES

A.   MAJOR CONSTITUENTS

Table I shows analytical techniques available for major constituents.

The classical techniques need no comment since only by refinement are they different from techniques used in school laboratories. The instrumental techniques for major con-stituents generally produce results that are reproducible and accurate to between $\pm$ 0.2 and $\pm$ 0.5% of the amount present. New techniques are continually being developed

and applied.   Microprobe analysis, for the determination
of constituents of small areas of the sample, and Auger
spectroscopy, for superficial components, are two such
techniques.

TABLE I

TECHNIQUE	LOWER LEVEL P.P.M.
Optical emission spark	10
Optical emission arc	0.01
X-ray fluorescence	50
Atomic absorption	0.1
U.V. absorption spectrophotometry	0.1
Polarography	0.1

B.   TRACE CONSTITUENTS

Table II lists analytical techniques for the determination of trace impurities.

TABLE II

CLASSICAL TECHNIQUES	RANGE (%)
Gravimetric	0.1 - 100
Volumetric	0.1 - 100
Electrolytic	0.1 - 100
INSTRUMENTAL TECHNIQUES	RANGE (%)
X-ray fluorescence	0.1 - 60
Atomic absorption	0.1 - 10
Optical emission spark	0.1 - 10
Optical emission hollow cathode	0.1 - 100

These levels of impurity are determined exclusively by
instrumental techniques.   The lower levels quoted are not

limits of detection for every element in any matrix, but a comparison of the lowest practical levels. Once again, new techniques are being developed to satisfy the increasing needs for more powerful tools for instrumental analysis. Mass spectrographs, particularly the newer instruments with photoelectric detectors, are capable of detecting concentrations of impurities much lower than can be determined by optical emission techniques. One problem with this method of analysis is that it is essentially a macular technique and the high sensitivity is derived from the low incidence and small diameter of the analytical spots. The small amount of sample removed for each analysis ($\sim$ 1 microgramme) makes calibration of standards in the sub parts per million level difficult. These instruments employ a high vacuum, and for this reason very few samples can be analysed in a day. Despite such limitations, the information provided has made possible some significant improvements in pure material technology.

Neutron activation is another technique which allows low impurity concentrations to be determined. The technique is, however, limited to impurities forming radioactive isotopes. Furthermore, the equipment is expensive and requires specialist facilities. Also transportation time is a critical parameter in methods using isotopes with particularly short half-lives. However, the advantage of this technique is that, unlike most of the other physical methods, it is a primary method and in the future may well be the reference method for trace analysis.

Optical emission spark techniques using metal or graphite electrodes can also be used for analysis. However, very few oxides are electrical conductors and the sample has

to be placed in an open cavity in the electrode.   The discharge can cause the sample to be ejected from the cavity
with a subsequent loss in sensitivity.   When the spark
technique can be controlled it works well, and excellent
agreement with chemical analyses has been obtained when
concentrations of the order of 1000 p.p.m. (0.1%) are
determined.

When comparatively high concentrations of impurity are
determined, photoelectric spectrographs (spectrometers)
provide a very rapid service, reporting 20 - 30 elements in
2 - 3 minutes.   Instruments such as these can be used to
determine very low levels of volatile impurities in oxide
materials.   The technique consists of condensing the vapour,
obtained by heating the sample, onto a copper finger which
is then used as an electrode in the system.   This can be an
effective concentration technique when the vapour from a
large sample is directed onto a small surface area.

Table II lists three methods which are seldom used for
the analysis of oxide materials, namely atomic absorption,
U.V. absorption spectrophotometry, and polarography.   The
reason for this lack of application is that in all these
methods of analysis a solution of the sample is used.
Oxides sometimes dissolve in water, but in the majority of
cases, strong concentrated mineral acid or fusion with
alkaline or acidic compounds is necessary.   The problem
then becomes the determination of the difference in impurity
levels, as the material used to effect dissolution will also
contain impurities.   The problem could well be to differ-
entiate between 10 and 10.01 p.p.m. when the preparation and
analysis of the dissolving medium becomes more important
than that of the sample.

C.   SPECTROGRAPHIC TECHNIQUE

The discussion so far indicates that the only viable techniques remaining are X-ray fluorescence and optical emission (arc) and these are the techniques used in the analysis of most oxide materials.   Because X-ray fluorescence has a comparatively high 'lower detection level', optical emission spectrography is preferred for qualitative and quantitative analysis of oxide materials.   Its advantages include speed, low analysis cost per sample, high sensitivity, the provision of a permanent record of those elements sought (and of others which may later turn out to be of interest), and a wide range of elements which may be estimated or determined.   The following elements can be determined by optical emission spectrography: Ag, Al, Ar, As, Au, B, Ba, Be, Bi, Br, C, Ca, Cd, Ce, Cl, Co, Cr, Cs, Cu, Dy, Er, Eu, F, Fe, Ga, Gd, Ge, He, Hf, Ho, I, In, Ir, K, Kr, La, Li, Lu, Mg, Mn, Mo, Na, Nb, Ne, Nd, Ni, O, Os, P, Pb, Pd, Pr, Pt, Ra, Rb, Re, Rh, Ru, Sb, Sc, Se, Si, Sm, Sn, Sr, Ta, Tb, Te, Th, Ti, Tl, Tm, U, V, W, Xe, Y, Yb, Zn, Zr.

The elements underlined cannot be determined by DC arc techniques, and even with these omissions the list is quite formidable.   The photographic plate provides a permanent record of those elements present in the sample which can be excited to produce light sufficiently sensitive to affect the emulsion.

Optical emission spectrography is a comprehensive technique and reliable analytical information may be obtained by attention to detail and an understanding of the processes involved in the procedure.

The process from sample to result can be divided in the four following parts: (1) excitation, (2) dispersion,

(3) integration,   (4) calibration.

1. Excitation

Sources of excitation are of three general types,
flame, arc, and spark.   For oxide materials the DC arc is
used since it does not depend on solution of the sample
(flame) or for the material to be electrically conducting
(spark).   With this energetic source of excitation very low
levels of impurity can be detected in a wide variety of
oxides.   The DC arc is a simple, comparatively inexpensive
and easily operated excitation source.   Quantitative
information of a very high standard may be obtained from the
DC arc if appropriate attention is paid to detail.   As a
rule, currents up to 50A are employed to obtain maximum
sensitivity, although some methods have been developed using
currents as high as 650A.   Increased current does not
necessarily mean improved sensitivity.

2. Dispersion

Dispersion is obtained by either a prism or a grating
according to the instruments used.   With increasing atomic
number there is a tendency for the spectra to become more
complex.   Consequently, the uranium spectrum when viewed
with a low dispersion prism spectrometer does not appear to
be a line spectrum.

Grating technology has now developed to an extent such
that very few instances are reported where an impurity
cannot be determined because of lack of resolution, whereas
many coincidence problems arise in prism instrument technology.

3.   Integration

Two main types of integrators are available for collection of the spectral light.   These are the photographic emulsion (film or glass plate) and the photo-multiplier.   Other methods are available but these have not generally been applied to spectrometric instruments.

The main advantages of the photographic plate are low cost and high information content.   Disadvantages are the non-linearity of response, intermittency and reciprocity failure and a tedious developing process.   Furthermore, the gelatin used in plate manufacture is subject to wide property variations.   However, by careful processing, good results can be obtained using the photographic plate as a detector.

For very specific applications the photographic emulsion can be replaced with a much cleaner, linear detector, namely the photomultiplier.   This device imposes many restrictions on the design of spectrometers and very specific techniques have to be employed when using this detector for trace analysis.

4. Calibration

Like all other physical methods this technique is comparative, and all measurements must be taken relative to some known standards.   Control of contamination and the heterogeneity of materials is most important in standard preparation.   Data processing, operation and parameter control of spectrographs is now commonly carried out by a computer.   Hitherto, experiments necessary to obtain an understanding of error functions, regression coefficients, limit theories and plasma kinetics were severely restricted or never performed because of the time and effort necessary to get the answers.

GENERAL INDEX

307